Primate Socialization

Primate Socialization

edited
by Frank E. Poirier

OHIO STATE UNIVERSITY

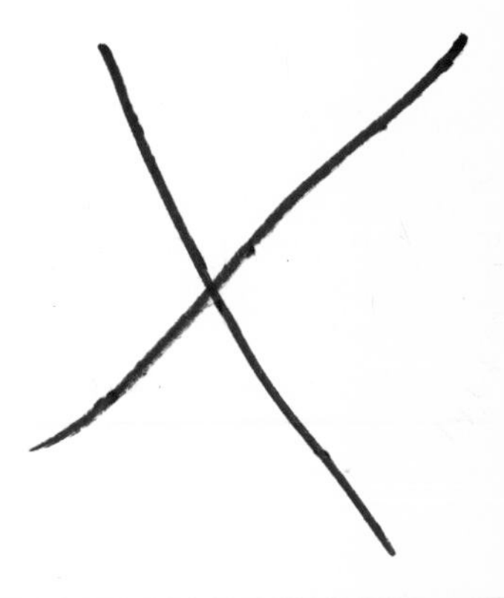

Random House New York

Published in the United States by Random House, Inc., New York, and simultaneously in Canada by Random House of Canada Limited, Toronto.

Library of Congress Cataloging in Publication Data

Poirier, Frank E. 1940–
Primate socialization.

Includes bibliographies.
1. Primates—Behavior. 2. Social behavior in animals. I. Title.
QL737.P9P64 599.8'045'24 78–173907
ISBN 0–394–31112–4

Typography by James M. Wall

Manufactured in the United States of America by Kingsport Press, Inc., Kingsport, Tenn.

First Edition

9 8 7 6 5 4 3 2 1

This book is dedicated to my wife Darlene,
to my daughters Alyson and Sevanne, and to my son
Shawn, who in ten short weeks of life taught me more
than I shall know in a
lifetime.

Preface

The contents of this book are an attempt to start a dialogue about a relatively unknown area of primate behavior—socialization. In order to do this, a number of investigators were contacted and asked if they would be willing to contribute. Some replied negatively, stating that their data were insufficient for this type of analysis. Those who have contributed are somewhat representative of the field of primatology, for example, the authors include scholars trained in England, Japan, and the United States. The scientific disciplines represented by these authors include anthropology, psychology, and zoology. Thus, the articles represent a cross section of thought and methodology.

It should be obvious to all who read this book that it is only a beginning. Future primatological research will answer some of the many questions raised in these selections. Perhaps this book will serve as a stimulus for future research on the topic of socialization.

The articles in this book cover a wide range of topics, which should be of interest to persons dealing with the topic of human and nonhuman socialization and behavior. The selections are diverse; however, they do not presume a store of background knowledge on the part of the reader. The data are presented in such a way as hopefully to be useful to student and professional alike.

Frank E. Poirier

COLUMBUS, OHIO
16 APRIL, 1971

Contents

Primate Socialization

Frank E. Poirier

Introduction

In this article I will be concerned generally with the socialization process of nonhuman primates. To simplify the data, the article is divided into subsections, each dealing with a specific feature of the socialization process. The work begins with a brief introduction. The second includes a brief review of the subsequent articles in the book and discusses the theoretical and practical limitations of our data. A theoretical framework for the socialization process is presented in the third section. Subsequent sections deal with the biological bases for nonhuman primate socialization and note several of the variables that impinge upon the process. In the fourth section, for example, we discuss such factors as the number of infants per birth, the retardation of growth, and nursing behavior. In the fifth section we note how the age and sex of the socializing agent(s) and of the animal(s) being socialized affect the process. The sixth deals with the mother-infant dyad, specifically the mother-infant feedback system, the factors upon which the mother-infant relationship rests, and the influence of the mother's behavior upon the behavioral development of her infant. The seventh section discusses the highly variable relationships between adult males and infants. The eighth deals with the role of play behavior in the socialization process. It is suggested that play behavior is a more important socializing agent in some species than in others. The ninth deals with the social structure as it impinges upon the socialization process. The species-specific social structure in which an infant is socialized influences who an infant will contact and how much contact time it will spend with which animals. In the tenth section we discuss the socialization process in two species, the terrestrial savanna-dwelling baboon and the semiterrestrial north Indian langur. This example illustrates how some of the variables previously discussed influence an infant's development. The article ends with a summary of the major points previously made.

THEORETICAL APPROACH

Numerous investigators, working in different situations and within divergent theoretical frameworks, have long reported the results of their research on the ecology and behavior of nonhuman primates. This information presently is accumulating at a rate faster than attempts to draw working hypotheses from it, although some workers (cf., Crook, 1967; Crook and Gartlan, 1966; Gartlan, 1968; and Kummer, 1967) have attempted to bring some order to the field studies by relating ecology to behavior and social organization. However, a wider theoretical framework seems necessary to understand the variables that contribute to the diversity of nonhuman primate societies and social behavior.

Since most primates live in highly complex, bisexual, year-around social groups, one fruitful avenue of research would be to attempt to discern how the socialization process—that is, how an animal's learned social relationships with its fellow social members—affects its behavior, and eventually, through such behavior, the social order. This work is a preliminary attempt to delve into the nature and consequences of this process.

Some colleagues caution that we lack the longitudinal data that will enable us to fully understand the nonhuman primate socialization process. While it is true that in many cases we lack longitudinal observational data, we do have enough information to begin an analysis and discussion. Each contributor to this volume fully agrees that his ideas are subject to change with further research. However, even granting the limited nature of some information, the efforts described here may help channel future data collection. To this end the discussions will have served a valuable function in our common concerns.

A REVIEW OF THE STUDIES AND THEIR THEORETICAL APPROACHES

Although there are voluminous data about nonhuman primates, only a small portion of the data are addressed, sometimes inadvertently, to the socialization process. The work of some Japanese investigators (cf., Itani and Sugiyama) approaches the ideal of a longitudinal study. Itani's data on the Japanese macaques draw on information that Japanese primatologists have collected during the past twenty years, mostly on provisioned troops.[1] Among provisioned Japanese macaque troops individual animals, as well as their genealogies, are recognized. This material has been subjected to analysis by numerous investigators familiar with the work. The chimpanzee material from the Budongo Forest, which Sugiyama reports, has been collected over several years by various investigators from the Japan Monkey Centre. Sugiyama's material can be compared with the

[1] Provisioned troops are troops of monkeys that are regularly fed by humans in specified areas. In the course of time, because the animals regularly return to the provisioning ground to eat, the investigators become familiar with them.

information recorded by Goodall (1965) on groups inhabiting the Gombe Stream Reserve. Chalmers, Lancaster, and Ransom and Rowell have reported on their studies of various African Cercopithecinae. Chalmers' data come from four species of enclosure-dwelling monkeys and are illuminating, since these four species normally inhabit divergent ecological niches. Vervets, for example, are predominantly terrestrial; De Brazza's monkeys inhabit the forest fringes and swamps and are both arboreal and terrestrial; Sykes monkeys are forest-dwelling arboreal animals; and mangabeys are highly arboreal monkeys. Since most of our data on baboons is derived from savanna-dwelling forms (DeVore, 1963; DeVore and Hall, 1965; Hall and DeVore, 1965; Washburn and DeVore, 1961) Ransom and Rowell's data on forest-dwelling groups provide comparative data that allow us to speculate about the specific effects of ecology upon the social order and behavioral maturation. Lancaster's information on vervets residing in the forests lining the Zambesi River can be usefully compared with data previously collected in other African econiches (cf., Brain, 1965; Hall and Gartlan, 1965; Gartlan and Brain, 1968; Struhsaker, 1967a, 1967b, 1967c, 1967d, 1969).

The emphasis in these studies is on wild or provisioned populations. However, wild or free-ranging studies impose limitations upon our analysis, for lengthy behavioral sequences may not be observed because of observational conditions. A full understanding of nonhuman primate behavioral ontogeny results only from continuous correlation and cross-checking of field and laboratory data. The total web of social learning, play and exploration, and sensorimotor coordination and the reflex system can only be described tentatively from observation of wild animals.

If the complex factors determining the behavior of mature animals living in social groups are to be discovered and evaluated, detailed experimental analysis seems to be required. We know from laboratory studies that the entire basis of group life can be eliminated by rearing animals in complete isolation (Harlow, 1960, 1962; Mason, 1960, 1961a, 1961b, 1963, 1965). We have learned that not only the form of social organization, but social life itself, is almost entirely dependent upon the individual's early rearing environment. Laboratory and field studies should supplement one another in experiments concerned with the development of nonhuman primate behavior. If possible, hypotheses about social learning derived from field observations of the socialization process in a primate group should be tested under controlled laboratory conditions. Laboratory observations of a few animals may shed light on some of the factors of social interaction and learning that are not immediately obvious when seen only within the context of the intricate structure of a wild social group.

Menzel (1967, 1968) and Mason (1968) noted some of the limitations of a free-ranging field observational situation on the study of the nonhuman primate socialization process. The field worker faces many uncertainties; for example, he takes a chance on finding a sufficient number of animals with which to work. More important for the purposes of studying the socialization process, under field conditions it is very difficult to complete a detailed analysis of a restricted developmental phase. To obtain such information it is often desirable to maintain relatively constant conditions. Special experimental testing is frequently required, and the animals must be constantly kept under close observation. These conditions can be most readily met under captive and laboratory conditions.

The articles in this text by Burton, Mitchell and Brandt, and Williams take a different approach to the nonhuman primate socialization process. Burton correlates various periods of social development with biological maturation. Her observation that a behavioral pattern such as sexual mounting derives from an infant's mounting a subadult shows that social behaviors may be predicated upon underlying species-characteristic biological/physiological processes. Furthermore, her observation that the appearance of some social behaviors can be influenced by the behavior of the leader male introduces a new variable into studies of the nonhuman primate socialization process. Mitchell and Brandt's article updates the former's work on paternalistic behavior (Mitchell, 1969). This material, compiled from many sources, is one of the few attempts to discern the male's role in the integration of youngsters into the social group. Many studies have neglected the male's socializing role (if indeed it exists in some species) because an infant's early and most obvious attachments are with its mother, and perhaps her previous offspring. Mitchell and Brandt's data help fill this gap. Finally, Williams's article discusses the general theoretical implications of nonhuman primate socialization and attempts to place this material in the context of human socialization theory. This article provides us with a wider theoretical framework for the other discussions in the text.

Each contributor has attempted to place his data into an ecological-phylogenetic framework so that the reader can appreciate the relationships between ecology, behavior, and social organization, when such relationships exist. The questions each author broached, for example, were "What makes a vervet a vervet and not a baboon?" and "By what processes of social interaction and behavior do nonhuman primates become viable group members?" Where possible, each author has attempted to show what roles specific group members played in the socialization process. In addition, some authors have attempted to define the ways in which various social behaviors, such as grooming and maternal behavior, help to integrate new members into the social group.

THEORETICAL FRAMEWORK

In this section we will establish a working definition of the socialization process. In order to do this we must first review the work dealing with the critical periods hypothesis. In addition, we will also discuss the results of the socialization process.

In a sense it is more difficult to define the socialization process and note its workings than it is to isolate its consequences. The major limitation of the theoretical framework of this text is imposed by the previously noted lack of longitudinal behavioral data. Much of our theoretical framework is therefore derived from socialization studies on humans (cf. Child, 1954; Clausen, 1968; Erikson, 1950; Freud, 1923, 1940), dogs (Scott, 1945, 1950, 1958), cats (Rosenblatt *et. al.,* 1961), rats (Collias, 1956), and sheep and other ungulates (Altmann, 1958), a highly diverse group of subjects.

Historically, the original models of the socialization process in nonhuman primates dealt primarily with the critical periods hypothesis expounded by Lorenz (1937), Scott and Marston (1950), Scott (1958), and Riesen (1961). Lorenz's 1937 work is the basis for the critical periods hypothesis. Lorenz found a circumscribed time in a young gosling's life when a moving object, whether goose or human, elicited a "following" response that remained stable throughout an animal's life. If not exposed to this stimulus during the specified time, a following response did not develop. The all-or-none nature of this process has been demonstrated in several species, although it has been shown that the time when imprinting occurs is subject to modifications in the laboratory. Scott and Marston (1950) used the phrase "critical period" to describe the results of a deprivation study with puppies. The importance of this study has since been questioned by Schneirla and Rosenblatt (1963). Riesen (1961) suggests substituting the phrase "critical stimulation"—for example, the exposure of an organism to specific stimulus complexities as it matures—which is most intimately associated with the acquisition of responses regardless of the stage of development. However, the subtle interaction between the physical developmental state of an organism and the stimulations to which it is exposed may have greater importance than suggested by Riesen.

The critical periods model espoused by Lorenz and Scott, while it may deal adequately with goslings, seems too simplistic an approach when dealing with nonhuman primates. There is, after all, no inherent reason why data collected from studies on cats, fish, ducks, chickens, rats, and so on, can be applied equally well to complex social animals. It is quite apparent that there are significant differences between such animals and nonhuman primates, differences that are exemplified in the complex social orders of the latter. The complex nonhuman primate social groupings add many dimensions not present in the development of goslings, for example. Studies of sensory functions and imprinting and studies of primate (including human) infants all suggest that the critical periods become much less fixed in time and more diffuse in nature as one moves across the phylogenetic scale.

The factors impinging upon the socialization of the nonhuman primate infant are complex and varied. The socialization process refers to an exceedingly wide range of complex phenomena. It simultaneously refers to the external stimuli received by an organism, the individual nature of the process, and to the end product or consequences of socialization. The term "socialization" will be used here to refer to the sum total of an individual's past social experiences, which, in turn, may be expected to shape its future social behavior. Since an individual is the outcome, the "product," of a given socialization process, it is appropriate to look at the variables influencing this output both vertically (in terms of time) and horizontally (in terms of social interactions). The product, or consequence, of the socialization process depends not only upon the original genetic material of the individual and the degree to which climate, nurturance, and other factors permit the realization of that potential; it is also influenced by the behavior of the adults and peers with whom the individual is or has been in regular contact.

Socialization can be viewed as that process which links an ongoing society to the new individual. Through socialization a group passes its social traditions and ways of life to succeeding generations. Socialization ensures that adaptive behavior will not have to be "rediscovered" anew in each generation.

THE BIOLOGICAL BASES OF SOCIALIZATION

The socialization of nonhuman primates is influenced by specific biological characteristics. Among these are the fact that primate mothers normally bear only one infant each birth, that much of the primate infant's early perceptions of the world are from the mother's stomach or back, that a primate mother nurses her infant, and finally that primates have a longer growth period. With the retardation of growth and a longer period of infant dependency, there is a clear tendency for individual experiences to play a more subtle role in shaping behavior into effective patterns. The extended period of nonhuman primate infant dependency enhances the amount and complexity of learning that is possible, and thus increases the adult's opportunities to shape the infant's behavior to meet local environmental conditions. Flexibility of behavioral patterns may be one of the principal benefits of the long nonhuman primate dependency period.

Learning now appears to be more important in the social development of nonhuman primates than many have acknowledged (cf., Mason, 1965; Washburn and Hamburg, 1965), since there appears to be a definite relationship between prolonged postnatal dependency and the increasing complexities of adult behavior and social relationships. A prolonged period of youth allows for learning in infant and juvenile play groups and provides more time for regular contact with adult group members. The importance of this is that contacts with adults other than the mother probably promote the socialization process and help to integrate young animals into the social group. Regardless of the fact that in many species such contact is limited, it may be of more importance than has been recognized to date.

VARIABLES INFLUENCING SOCIALIZATION

In this section we discuss how the age and sex of both the socializing agent and the individual being socialized affect the socialization process. Another variable, the nature of interanimal contacts, is also discussed.

The primate growth cycle may be divided into four periods (Harlow, 1966; Scott and Marston, 1950). Since socialization is a continuous process occurring at all stages of development, it may seem unusual to speak of age as a variable. However, it is now clear that the importance of any specific social event may vary according to the age at which the individual experiences it. The specific or the average age of an individual may be significant as an antecedent variable in the socialization process in the following ways: (1) There may be a simple causal relationship between age and some dependent variable. It is possible that in some nonhuman primate species early socialization has more pronounced effects than later socialization. (2) There may be a more complex effect of age, not to be expressed in simple quantitative terms. For example, the socialization process may have qualitatively different kinds of effects at various periods in the life cycle.

The first period in the nonhuman primate life cycle may be designated the *neonatal period.* This is the time when locomotor and ingestive behavioral patterns are suitable only for infantile life. During the neonatal period mother-infant contact is continuous, close, and of a long duration. The importance of this period in the socialization process seems directly related to the amount and kind of maternal protection and solicitude available to the infant. The second period may be labeled a *period of transition,* during which time adult locomotor and ingestive patterns overlap, at least in unskilled manifestations, infantile forms. This period terminates when the young leave the regular company of their mothers. In primates, the transitional period is gradual, extending over several months. Considering the animal's total life span, the length of time in this period is relatively short. The third period is a time of *peer socialization,* when the youngster contacts individuals other than its mother. These contacts primarily involve the mother's prior offspring, older females, and age mates. This period is characterized by the gradual subsidence of infantile patterns of behavior and is marked primarily by the weaning process. The fourth period, the period prior to the assumption of full adult participation in the social order, can be termed the *juvenile/subadult period.* During this stage infantile patterns disappear and adult behavior patterns (such as sexual behavior) are practiced. The length of the *juvenile/subadult* period depends upon the sex of the animal (nonhuman primate females tend to become full adults earlier than the males) and the longevity of the species.

For each period in the life cycle we can define four main elements in the socialization process: (1) the typical life condition, or style, dominating the attention of the adults and their offspring; (2) the agents of socialization, those individuals and social units that typically play a role in the socialization process in each of the several stages of the life cycle; (3) objectives that these agents may follow or set as goals; and (4) the main learning tasks facing the individual, the problems to be solved or the skills to be learned as they confront the individual from its internal personal perspective.

Beyond the age variable, other influences must be recognized. These include the various types of interanimal relationships. In human terms, Freudian theory stresses the importance for personality development of a child's social relationships with other family members, and by implication suggests that the consequences of this interaction may be greatly influenced by the character of other members as socializing agents. General behavioral theory, as applied to humans, views personality development as importantly influenced by learning in social relations. It also suggests that what is learned depends upon the agents of socialization. Some of the variation among different socializing agents is a matter of individual idiosyncrasies, but some may be associated with social status characteristics. The socializing effects of an adult may differ greatly from those of socialization by peers and elder siblings. Interactions with one's parents may have different results from similar interactions with adult nonrelatives.

The sex of the socializing agent also makes a difference. The several variables defined for the human situation, that is, *sex, age,* and *adulthood status* of the socializing agent, and the *degree of social intimacy,* all seem to also influence the socialization process of nonhuman primates. Such variables have received inadequate attention in nonhuman primate field studies.

THE MOTHER-INFANT DYAD: AN EARLY BASIS OF SOCIALIZATION

Of all the relationships that a youngster maintains, its ties with its mother are the earliest, strongest, and perhaps the longest lasting. Many observers feel that the nature of the mother-infant relationship influences the infant's course of development and social contacts. This section discusses the mother-infant dyad generally and in the context of specific social structures. We will also discuss the influence of the mother's social status upon the social status of her infant. Finally, there is a brief discussion of the weaning process, and indications of how variations in this process may affect an infant's behavioral development.

Of the various individuals, or combinations of individuals, affecting the socialization of an infant, the following deserve attention: the mother-infant affectional system, the peer affectional system, the heterosexual bond, and the bond between older and younger animals (Harlow, 1966). The mother-infant dyad has received the most attention in recent studies. It has been suggested that all the major social roles and classes of bonds (i.e., male and female, dominant and subordinate) may ultimately have their roots in the initial socialization of the infant by its mother. However, this point remains to be validated in field and laboratory studies.

The maternal relationship is the first affectional bond for the youngster (Harlow, 1962, 1963) and is perhaps the prototype of all later such bonds (Jensen *et al.,* 1967). Possibly, as clinicians relate for humans, the relationship between older animals simply is an elaboration and a redirection of the initial attachment to the mother (Jolly, 1966). There seem to be notable differences between the ways various nonhuman primate mothers handle their infants and in the amount of time they are in physical contact with them (Jay, 1962; Poirier, 1968; Tinklepaugh and Hartman, 1932, among others). Such differences may result in the wide spectrum of adult behaviors reported in the literature. For example, the fact that some primates at the same phyletic level lead solitary lives while others live in groups raises the question of whether these opposed tendencies are inherited or acquired. (The range of social organizations among lemurs is illustrative; Jolly, 1966.)

For all primates the mother is not only the central feature of the social and physical environment, she also serves as the infant's locomotor organ and neocortex and so determines the nature of the basic socialization environment. The neonatal primate clinging to a mobile mother forms an attachment not only to her per se but through her to virtually her whole ecological-social setting. Later attachments may simply be differentiations and specializations of this early and relatively amorphous monolithic state.[2]

In neonate mammals behavior from birth is typified by reciprocal stimulation between parent (more specifically, mother) and offspring. The infant activates the female's stimulation, she in turn presents the newborn with a variety of tactile, thermal, and other stimuli, typically of low intensity and primarily approach-provoking. On this basis, the socialization process commences. Thus, behavioral development is essentially social from the beginning. The principal factors in this development are those involving the perceptual development of the female, processes of individual perception of the young, and the reciprocal stimulative relationships between a female and her young.

With the formation of an attachment to the mother, favorable conditions prevail for social learning, for it is evident that social learning begins with the mother. The occasions for such learning are multiplied as the growing animal moves from its mother to others. Experimental investigations have paid little attention to the task of specifying the manner in which this experience contributes to normal social development. However, we can assume that one of the primary functions of contacts with other animals is to sharpen, strengthen, or generalize the learned behavior originating in the mother-infant relationship.

The species-specific social milieu in which the nonhuman primate maternal-infant relation develops exerts a prominent behavioral force (Mason, 1965). A noticeable contrast appears in the langur versus the baboon-macaque pattern of maternal care. Unlike most monkeys, langur mothers allow other females to hold and carry a newborn. North Indian langur mothers begin to pass their infants to other females a few hours after birth, as soon as the infant is dry. "As many as eight females may

[2] The young of most mammals, relatively helpless at birth, experience prolonged periods of dependence upon parental sources of nourishment and protection, and they learn to make social accommodations according to the behavioral modes of the mother. The young form habits of behaving with and in conformity to the behavior of the parents—in the case of the nonhuman primates, the mother and others of their kind. Although physiological and morphological states influence the nature and extent of the early child-parent relationship, psychological factors and social habits formed during infancy (and perhaps slightly later) influence the nature and extent of social tendencies that persevere later (Tinklepaugh, 1948).

hold the infant during the first day of life, and it may be carried as far as 50 feet from the mother" (Jay, 1965b, p. 221). There are variations on this theme among langurs, however. North Indian and Nilgiri langurs (Poirier, 1968) willingly give up their infants (and their infants seldom resist leaving), whereas the south Indian langur infant "resists other females, clinging to the mother's body and squealing loudly and violently" (Sugiyama, 1967, p. 228). In ten of forty-nine infant transfers witnessed among the Malaysian form *P. cristatus,* the infant vigorously resisted by clinging to the original female and attempting to run from the second female (Bernstein, 1968).

The macaque and baboon mother-infant relationships differ. Baboons are attracted to a newborn infant, and soon after birth an animal may approach to touch or to groom the infant. They do not attempt to take the infant from the mother, however, nor does the mother leave her infant with others (DeVore and Hall, 1965). A bonnet macaque mother does not allow other females to hold her infant (Simonds, 1965). A rhesus macaque mother is not the center of attraction of other females or infants, and the infant has limited contact with group members other than its mother during the first weeks of life (Southwick *et al.,* 1965). In stump-tail macaques (Bertrand, 1969), however, adult and subadult females are strongly attracted to young infants, some of the females even acting as a "protective aunt," a role that Rowell and Spencer-Booth (1964) found important among captive rhesus macaques.

ACQUISITION OF SOCIAL STATUS

A brief look at the origin of social status is illustrative, for the few reports available from long-term studies on macaques suggest that a mother's rank influences her infant's status. This is accomplished primarily through the infant's mimicking of its mother's social interactions with other adults and by the infant's ability or inability to count on the mother's support in case of trouble. The concept of identification has been introduced from psychoanalysis into studies of nonhuman primates to explain the fact that Japanese macaque infants with dominant mothers tend themselves toward dominance (Imanishi, 1957). Infants of higher-ranking mothers had substantial contact with and identified successfully with troop leaders. Offspring of lower-ranking mothers had minimal, if any, contact with troop leaders and were unable to identify with them during childhood. In the Takasakiyama troop they became peripheral members or deserted the social group (Itani *et al.,* 1963). Studies of the Koshima (Kawai, 1958) and Minoo B (Kawamura, 1958) macaque troops in Japan show that in paired competition for food, successful monkeys were often infants of higher-ranking mothers. Koford's (1963) and Sade's (1965, 1967) reports on the Cayo Santiago rhesus macaques indicate that the adolescent sons of the highest-ranking females hold a high rank in the adult male structure. Baboon infants of lower-ranking mothers exhibited considerable insecurity in the form of a greater frequency of alarm cries and more demands on the mother, leading to an intensification of the mother-infant bond. Offspring of higher-ranking females, however, acted more secure and exhibited more freedom from their mother (DeVore, 1963).

A different situation obtains among Nilgiri langurs (Poirier, 1970). Dominance is not an important feature in the life of an adult female Nilgiri langur and her status, or social position, was seldom apparent in the daily course of events. There is no indication that her status had any measurable effect upon her infant's social behavior. Most likely, a female's dominance status was less influential in terms of the social development and dominance status of the infant than was her "temperament," which may have affected the total pattern of her maternal behavior. Every Nilgiri langur infant has free access to every other infant; females of all ranks have free access to all infants (Poirier, 1968). The Nilgiri langur infant was reared in a different environment and by a different process than obtains among the macaque and baboon.

WEANING BEHAVIOR

One aspect of the nonhuman primate maternal-infant behavioral system that is largely ignored, except on a gross level, is the weaning process. During weaning

> . . . there is a gradual transition of the dyadic relationship in monkeys and apes from an initial period of virtually continuous physical attachment and co-directed attention, through several transitional stages, to an ultimate stage of independent and separate functioning of the offspring and mother. Such independence is obviously necessary for the infants to enter into the adult activities of their species, and for the mother to turn her attention to the next offspring when it comes [Kaufman and Rosenblum, 1969].

The weaning process encompasses the physical and emotional rejection by the mother of the infant. The mother, once the major source of comfort, warmth, and food, now is hostile and denying. The severity of this rejection seems to depend upon the temperament of both the mother and the infant. Some mothers reject their infants more positively than others, and some infants are more persistent in their attempts to resist rejection. Older females who have borne many young probably wean their infants with less effort than younger females (Jay, 1962).

As early as 1935 Mead, in her book *Sex and Temperament in Three Primitive Societies,* remarked on the association of adult aggressiveness with rough and abbreviated nursing habits in contrast to the gentle and prolonged nursing in a more cooperative tribe. Heath, in an unpublished experiment on early weaning and aggressiveness in rats, found a significantly greater degree of aggressiveness in nine early weaned rats as compared to nine young rats permitted to remain with the mother. Most assuredly there is a wide range of weaning behavior among nonhuman primates, not only in procedure but in moment of onset, and pursuit of this line of research might be interesting. Some laboratory and field reports suggest that there are sexual differences during the weaning process. Among laboratory groups of pig-tailed macaques, *M. nemestrina,* male infants left the mother more often than female infants, and mothers left male infants more than they left female infants (Jensen *et al.,* 1967). Reports on wild Japanese macaques are consistent with these data (Itani, 1959). Itani notes that infant males left their mothers to form male peer groups at an age when young females remained with their mothers. In a wild study of *P. johnii* (Poirier, 1968), no sexual differences in weaning behavior were noted. If lack of sexual differences in weaning among Nilgiri langurs reflects a true situation rather than problems with data analysis, perhaps the greater behavioral variation noted, for example, among Japanese macaque males and females as compared to Nilgiri langur males and females is somehow related to this early developmental period.

Anthoney (1968) suggests that there is an ontogenetic development of grooming stemming from nursing and weaning behavior. He notes that grooming first becomes important for the infant when it is weaned from the breast. Although the mother disallows nursing, she usually tolerates the infant's attempts to groom her. Thus, whenever the infant is frightened or otherwise needs security, the weaned infant comes to groom rather than nurse. Given this as a possibility, there could be some link between the amount of grooming in a species and the length of the nursing period.

INFANT–ADULT MALE RELATIONSHIPS

Perhaps the most variable relationship, in terms of amount of contact and time of onset, a youngster has is that with the adult males of the group. In some species the infant–adult male relationship is minimal. The Nilgiri langur is used as an example of how the social structure minimizes the amount of infant–adult male contact, and the Japanese macaque as an example of how the social structure allows some infants contact with dominant adult males and other infants no contact with them.

As the infant matures, part of the behavioral repertoire developed during interactions between the mother and infant is extended to other group members. The infant's social world rapidly expands to include more of the group, first embracing its peers. Much of the earlier social contact is made on the infant's initiative, and infants of one species may not contact adults until a later age than do infants of another species. The greatest variability in infant contact is with the adult males. In some groups, such as north Indian langurs (Jay, 1965b), this primary contact is ritualized; in others it just "happens," so to speak. In some species, for example, hamadryas baboons (Kummer, 1968), subadult males adopt infant females and thus form the basis of their own group. In Gibraltar macaques (Burton, this volume; MacRoberts, 1970) subadult males play a role in the integration of the infant into the group. Among Nilgiri langurs (Poirier, 1970) males seem to have little or no role in the socialization process. The amount and the time of onset of infant–adult male contact must affect the socialization process.

The troop's social structure largely determines the amount and type of infant–adult male contact. This may be illustrated by comparing Nilgiri langurs (Poirier, 1969) and Japanese macaques. Nilgiri langur groups are organized into fairly consistent subgroups, that is, assemblages of animals having a recognizable affinity from day to day. Each subgroup is a social aggregate of individuals of similar age and/or sex. Most social interaction occurs within rather than between subgroups. Adult males and adult females rarely interact socially with one another. Thus, Nilgiri langur infants and juveniles rarely interact with adult males (especially when females in the subgroup give birth). Japanese macaque troops, on the other hand, are comprised of central and peripheral portions. Infants born in the central part of the troop identify with troop leaders; they in turn tend to become leaders. Whereas Nilgiri langur infants all go through a similar rearing process, some Japanese macaque infants receive special attention from leader males, passing through a different rearing process than do infants on the troop's periphery.

THE ROLE OF PLAY BEHAVIOR IN THE SOCIALIZATION PROCESS

In this section we will discuss the role of play behavior in the socialization process. Although there is some difficulty defining exactly what play is, most observers agree that play does help socialize the youngster.[3] Behav-

[3] Although there is no one definition of play behavior that satisfies all observers, most do agree upon a play sequence when one occurs. Loizos (1968) defines play as a positive approach toward and a nonrigidified interaction with any environmental feature, including group members, which involves stimulation through most sensory

ioral patterns common to sexual and dominance relationships are practiced in the play groups. An animal may learn its position in the dominance hierarchy during play-fighting bouts. It is possible, however, that play is a more important socializing agent in some species than in others. Terrestrial primates seem to spend more time in play activities than do arboreal species.

Much of the nonhuman primate infant's earliest contacts beyond the mother are with its peers; a large portion of this early contact occurs as play behavior. Play is one of the first non-mother-directed activities appearing in nonhuman primate behavioral development. Mason (1965, p. 530) notes that "playfulness . . . is rightly regarded as a useful index of the physical and psychological well-being of the young primate. Its prolonged absence raised the suspicion of retardation, illness or distress."

Infants seem to adjust to their fellows and learn to become effective members of society with the help of play behavior. Through trial and error, through the constant repetition of behaviors characteristic of play, an infant learns the limits of its self-assertive capabilities. The play group is a context for such learning because its members, mostly peers, are young and their teeth are neither sharp nor long enough to inflict damage. The adult dominance hierarchy may form its basis in the play group wherein youngsters compete for food, sleeping positions, or the easiest arboreal pathways. Although aggressiveness seems to derive from the play-fighting characteristic of young animals, certain forms of play may help condition the young to cooperative, group-positive behavior (see Lancaster, this volume). In the play group the juveniles establish the close social bonds that will later help maintain group unity. Infants learn to mix within the play context; by playing with their peers they develop fully integrated personalities.

It has been suggested that the period of maximal social play that occurs in nonhuman primates corresponds to the brief period of imprinting in birds. During the play period the primate youngster becomes familiar with the sights, sounds, smells, and touch of its inanimate and animate environment. During play the animal learns which species it belongs to. Far from being a "spare-time," superfluous act, play behavior at certain crucial early life stages seems necessary for the occurrence and success of all later social activities. The very quality of play (exaggeration, or lack of economy, in movement) ensures maximal energy expenditure, thus in-

---

modalities. The behavioral patterns characteristic of play are motorically similar to those occurring in contexts where they serve immediate and specific biological functions. However, they differ qualitatively from these patterns in that movements are exaggerated and uneconomical; they would be inefficient in terms of the originally motivated context.

creasing the strength of the social learning process. The longer the period of dependency in infants, the more vulnerable they are to influences and interactions that affect their social development adversely, and therefore the more essential it is that there be a means of ensuring continuous and corrective interactions with their species. Play might serve this function (Loizos, 1968).

In many species social play between subadults occupies much of the time that is not spent in eating and sleeping. The patterns of social play are largely derived from those of agonistic behavior, consisting of chasing, wrestling, tumbling, biting and dragging, or chewing. Adjustments appear during play that enable a young primate to function properly as an adult member of the species and to occupy a position within the social organization of its group. It is generally agreed that social play contributes to the behavioral maturation of the young. Youngsters must have an opportunity for adequate play sessions, for if they do not, they are faced with the options either of being maladjusted or of being excluded from the group (Carpenter, 1965).

The importance of play for normal social development has been adequately demonstrated in laboratory situations. Even brief daily play sessions between infants raised by surrogate mothers fully compensated for their lack of real mothers. At similar chronological ages, these infants developed as complete a repertoire of infant-infant play relations, and later on, adult sexual relations, as did infants raised with their mothers in playpens. Surrogate-reared infants allowed twenty minutes of play daily with their peer group were considerably better adjusted (as adults) than infants raised with mothers alone. There is strong supporting clinical evidence that interactions with the peer group are both necessary and sufficient for the development of normal adult social behavior.

Beach (1945) suggested that there is a relationship between phylogenetic position and the amount, duration, and diversity of play behavior. However, the relationship may not be direct, as suggested by Loizos (1968). Lorenz (1956) and Morris (1964) suggest a distinction between animals whose mode of survival is highly specialized, structurally or behaviorally, and those who are "opportunists." The latter's distinguishing characteristic is their restless curiosity, what Morris calls their "neophilia," or love of the new. These animals generally maintain a higher level of activity than more specialized species. Of such animals, the primates are the supreme examples, the chimpanzee being the apex, as is demonstrated by its many and varied play behaviors.

Even though play behavior is an important element in a young primate's life, there is a great deal of variability between the amount of play and the participants that does not reflect solely phylogenetic or ecological variables. Nilgiri langurs, for example, are not very playful (Poirier, 1969, 1970). A total of 180 play sequences, or one play sequence per 6.9 hours of observation, were noted during 1,250 hours of field recording. In contrast to the minimal play behavior noted among Nilgiri langurs and substantiated by Tanaka (1965) for the Nilgiri langur groups that he studied, north and south Indian langur youngsters spent considerable time in vigorous play behavior. Macaques (Simonds, 1965) and baboons (Hall and DeVore, 1965) also indulge in a good deal of play activity.

It is possible that the limited number of playmates available in many Nilgiri langur groups influenced behavioral development. A maturing youngster in an average Nilgiri langur group unquestionably had fewer opportunities for peer social interaction than did the common langur, baboon, and macaque infant maturing in troops containing fifteen or twenty young playmates. Perhaps play behavior serves a less important function in the behavioral development of Nilgiri langurs than it does for common langurs, baboons, and macaques.

THE INFLUENCE OF THE SOCIAL STRUCTURE UPON THE SOCIALIZATION PROCESS

In this section we will outline how the social structure impinges upon the socialization process. Nonhuman primates are group-living, social animals; therefore, to fully understand the socialization process, it must be viewed within the context of the social structure.

If we suppose that the development of nonhuman primate infant behaviors involves the interaction of a genetically determined base with a set of environmental conditions, we must remember that when we examine primate infants, the environmental condition is largely social. Most primates are social animals living in groups, the structure of which reflects and influences individual behaviors (e.g., Poirier, 1969). From laboratory studies we know that the entire basis of group life is almost obliterated by rearing animals in isolation. To this extent not only the form of the group organization but group life as such is completely dependent upon the early environment of individual animals. Presumably, any given form of social organization is sustained as an adaptation to the ecology, but socialization impinges upon this process. Social structure intrudes upon, and in many ways influences, socialization. The reciprocal relationship between socialization and social structure is not necessarily one of discrete interactions but may take the form of cycles or other sequences prolonged over substantial time periods. The characteristics of primate social groups differ according to many variables, among which are sociality and the incidence of dominance, sexual, and grooming behavior. But, with few exceptions, higher primates live within a social group in which the animal learns to use its biology efficiently and adapt to its environment. The differences in primate societies are due not only to biology but to the social circumstances in which the individual lives and learns.

The species-specific troop social structure determines which animals an infant will or will not contact. For example, the subgroup assemblages characteristic of Nilgiri langurs allow an infant an inordinate amount of contact with its mother, other females, and peers, and little contact with the adult male (or males in multiple-male groups). In a previous work (1969) I attempted to show how some of the social differences characteristic of Indian leaf-eating monkeys are partially due to the amount of contact animals have with other specific individuals.

It is within the social group, replete with all its variations, that the young primate is socialized, where it learns to perform effectively as an adult member of the species. Mason's (1960, 1961a, 1961b, 1963) studies on rhesus macaques aptly demonstrate the effect of environment and social restriction on the maturation process. Animals with restricted social experiences, that is, those raised in isolation, show strikingly abnormal patterns of sexual, grooming, and aggressive behavior. Monkeys maturing in a restricted environment groom infrequently and are more aggressive than animals raised in a natural, wild condition.

Restricted monkeys avoid wild members of their species, and wild animals raised in normal situations fail to seek out animals reared in restricted conditions. Mason's (1963, p. 169) studies ". . . indicate . . . that our socially deprived monkeys were attracted to some feral animals, but that they were not attractive companions to each other, nor to monkeys born in the field." The full development of an animal's biological potentialities seems to require the stimulus and direction of social forces that are usually provided in the social group (Washburn and Hamburg, 1965).

TWO EXAMPLES OF CONTRASTING MODES OF SOCIALIZATION

In this final section we will outline the socialization process of the terrestrial savanna-dwelling baboon and compare this to the semiterrestrial north Indian langur. The objective is to note how variations in such things as maternal-infant behavior, infant passing, infant–adult male contact, and play behavior may affect behavioral development. These variations must be seen, however, in the context of different ecological adaptations and pressures.

Eimerl and DeVore (1965) have compared the social systems and the socialization process of langurs and baboons. Although both langurs and baboons live in complex, year-around social groups important differences exist. Baboon groups are considerably larger, sexual dimorphism is more pronounced, and there are no solitary or peripheral-group individuals. While the patterns of physical and social development are similar, the resulting adult behaviors differ, which acknowledgment of different ecological pressures does not fully explain. These differences become clarified if we look at the unique occurrences and experiences that influence the lives of baboon and langur infants, for example, that variations in the mother's reproductive physiology and temperament, and the youngster's experience with previous offspring and with members of the mother's lineage (see, for example, the work of Imanishi, 1957 and Kawamura, 1958 on Japanese macaques).

There is considerable variation in the manner in which langur and baboon mothers handle their infants. A newborn langur's experiences with its mother vastly differ from those of a baboon. Baboon infants usually do not leave their mother's arms during the first month of life, and baboon mothers are reticent about allowing other females to touch their offspring. But, the newborn langur is passed about to other females (to varying degrees among different langur species, Poirier, 1968, 1970). There is also a striking difference between the infant's relationships with the adult males. For baboons the relationship is very close, whereas for the langur infant there is little contact with the adult males until about eight months of age, and female langurs have almost no contact with adult males until they are about three years old. Also, it is the adult female langur that protects the infant, even chasing away males who may startle the youngster.

After weaning, infant langurs become segregated by sexes. Juvenile females stay near the group's center close to the adults, mixing more and more intimately with the adult females and infants. The juvenile and subadult females hold the infants and sometimes tend to them when the mother is gone. Juvenile male langurs, meanwhile, spend most of their free time playing. As they mature play becomes rougher, and needing more room to play, they drift to the periphery of the group, away from the infants and adults. Thus, they form a rather distinct subgroup.

The baboon-macaque pattern differs markedly. While the infant langur matures in almost a matriarchal society, baboon and macaque infants mature amid a mixed society of males and females. Unlike langurs, baboon and macaque males are interested in the infants; adult male baboons frequently approach the mother, smacking their lips to indicate no harm, in order to approximate the infant. Young male baboons from the age of nine months tend to have minimal contact with the females, who respond to them less and less as they mature. They get rough treatment and are often smacked harder and harder by the females, and eventually they leave their protection. Thus, the differences between langurs, baboons, and macaques, which are ultimately adaptations to different ecological niches, are tied to the socialization process.

SUMMARY

In this article we have attempted to set the stage for the following articles by defining the socialization process and by noting the numerous variables impinging upon it. In the broadest sense, socialization refers to the sum total of an individual's past experiences, which, in turn, may be expected to help shape its future social behavior. The socialization process can be viewed as a bridge; socialization is that process linking society to the individual. Through socialization a group passes its traditions to succeeding generations. Socialization provides a sense of continuity and ensures that adaptive behavior will not have to be "rediscovered" anew each generation.

The nonhuman primate socialization process is influenced by biological characteristics common to nonhuman primates. Basic among these is the fact that a female bears one infant at a time to which she can give almost undivided attention. Another important factor is that primates are characterized by an extended growth period, thus providing them with a longer period prior to becoming adults in which to learn behaviors applicable to adult life. Flexibility of behavioral patterns may be one result of this long growth period. There also seems to be a relationship between the prolonged period of postnatal dependency and the increasing complexities of adult behavior and social relationships.

Of the following relationships, the mother-infant bond, the peer bond, the heterosexual bond, and the bond between older and younger animals, the mother-infant bond has received the most attention. It is possible that all major social roles and types of social bonds ultimately have their base in the initial socialization of the infant by the mother. The amount of time a mother spends with her infant and the manner in which she spends this time may influence the youngster's behavioral development. Possibly, as clinicians relate for the human situation, the relationship between older animals is simply an elaboration and redirection of the initial attachment of the infant to the mother. The diverse character of the maternal-infant relationship found among nonhuman primates may be related to the wide spectrum of social organizations and adult behavioral patterns that are recorded in the literature. There is also the possibility that the manner in which a mother weans her infant is as important as the manner in which she treats it during the nursing period.

Besides the nature of the mother-infant dyad other variables are important. As the infant matures it extends its contacts to other group members. Its first prolonged contacts with animals other than its mother are usually with its siblings and peers. The amount and nature of the sibling relationship is yet to be adequately analyzed, but apparently it differs widely among the primates. Peer relationships, however, almost universally occur within the context of play behavior. Play behavior, as laboratory studies aptly demonstrate, is very important for the social and physical development of the youngster. The locomotor patterns common to play behavior strengthen the youngster's muscles and prepare it for adult life. Furthermore, play behavior helps the infant adjust to its social setting and helps it learn to become an effective member of society. Through trial and error, through the continual repetition of play behaviors, the infant learns the limits of its self-assertive capabilities and the characteristics of its environment. It has been suggested that the basis of the dominance hierarchy is established in the play group wherein youngsters compete for food, travel routes, and resting places. Within the play group youngsters may learn cooperative, group-positive behaviors, thus establishing close social bonds that will later help maintain group unity. Adjustments seem to appear during play that enable a young primate to function properly as an adult member of the species and to occupy a position within the social organization of the group.

If, as we suppose, the development of infant behaviors involves the interaction of a genetically determined base with a set of environmental conditions, then we must remember that the environmental condition of the nonhuman primate is basically social. When we survey the range of social organizations characteristic of the nonhuman primates, it should be evident that group social structure impinges upon the socialization process. For example, among Nilgiri langurs, which are characterized by rather rigid subgroups based upon age and sex, infants have more contact during their early years with their mothers and other females than they do with the adult males, who remain both physically and socially on the group's periphery.

In an attempt to elucidate how some of the variables discussed in the article influence behavioral and social development, we have compared the social order of the terrestrial savanna-dwelling baboon to that of the semiterrestrial north Indian langur.[4]

[4] Note: After the final version of this chapter was written, I read an article by J. H. Crook, "The Socio-ecology of Primates" in J. H. Crook (ed.), *Social Behavior in Birds and Mammals* (1970). This article deals with, and expands, many of the points included in Crook's article.

References

ALTMANN, M. "Social Integration of the Moose Calf." *Anim. Behav.* 6 (1958), 155–159.

ANTHONEY, T. R. "The Ontogeny of Greeting, Grooming, and Sexual Motor Patterns in Captive Baboons (Superspecies *Papio Cynocephalus*)." *Behavior* 31 (1968), 358–372.

BEACH, F. A. "Current Concepts of Play in Animals." *The American Naturalist* 79 (1945), 523–541.

BERNSTEIN, I. S. "The Lutong of Kuala Selangor." *Behavior* 14 (1968), 136–163.

BERTRAND, M. *The Behavioral Repertoire of the Stumptail Macaque.* Basel: S. Karger, 1969.

BETTELHEIM, B. *The Empty Fortress.* New York: Free Press, 1967.

BRAIN, C. K. "Observations on the Behavior of Vervet Monkeys." *Zoologica Africana* (1965), 13–27.

CARPENTER, C. R. *Naturalistic Behavior of Nonhuman Primates.* University Park: Pennsylvania State University Press, 1965.

CHILD, I. L. "Socialization." In G. Lindzey (ed.), *Handbook of Social Psychology,* Vol. II. Reading, Mass.: Addison-Wesley, 1954, 655–692.

CLAUSEN, G. A. (ed.). *Socialization and Society.* Boston: Little, Brown, 1968.

COHEN, Y. A. *The Transition from Childhood to Adolescence.* Chicago: Aldine, 1964.

COLLIAS, N. E. "Socialization in Sheep and Goats." *Ecology* 37 (1956), 228–239.

CROOK, J. H. "Evolutionary Change in Primate Societies." *Science Journal* (1967), 1–7.

——— and J. S. GARTLAN. "The Evolution of Primate Societies." *Nature* 210 (1966), 1200–1203.

DEVORE, I. "Comparative Ecology and Behavior of Monkeys and Apes." In S. L. Washburn (ed.), *Classification and Human Evolution.* Chicago: Aldine, 1963, 301–319.

——— (ed.). *Primate Behavior: Field Studies of Monkeys and Apes.* New York: Holt, Rinehart and Winston, 1965.

——— and K. R. L. HALL. "Baboon Ecology." In I. DeVore (ed.), *Primate Behavior: Field Studies of Monkeys and Apes.* New York: Holt, Rinehart and Winston, 1965, 20–52.

EIMERL, S. and I. DEVORE. *The Primates.* New York: Time-Life Books, 1965.

ELLSIN, F. *The Child and Society: The Process of Socialization.* New York: Random House, 1960.

ERIKSON, E. H. *Childhood and Society.* New York: Norton, 1950.

FREUD, S. *The Ego and the Id.* London: Hogarth, 1923.

———. *An Outline of Psychoanalysis.* London: Hogarth, 1940.

GARTLAN, J. S. "Structure and Function in Primate Society." *Folia Primat.* 8 (1968), 89–120.

——— and C. K. BRAIN. "Ecology and Social Variability in *Cercopithecus aethiops* and *C. mitis.*" In P. Jay (ed.), *Primates: Studies in Adaptation and Variability.* New York: Holt, Rinehart and Winston, 1968, 253–292.

GOODALL, J. "Chimpanzees of the Gombe Stream Reserve." In I. DeVore (ed.), *Primate Behavior: Field Studies of Monkeys and Apes.* New York: Holt, Rinehart and Winston, 1965, 425–473.

GOSLIN, D. (ed.). *Handbook of Socialization Theory and Research.* Chicago: Rand McNally, 1969.

HALL, K. R. L. and I. DEVORE. "Baboon Social Behavior." In I. DeVore (ed.), *Primate Behavior: Field Studies of Monkeys and Apes.* New York: Holt, Rinehart and Winston, 1965, 53–111.

HALL, K. R. L. and J. GARTLAN. "Ecology and Behavior of the Vervet Monkey, *Cercopithecus aethiops,* Lolui Island, Lake Victoria." *Proc. Zool. Soc. Lond.* 145 (1965), 37–56.

HARLOW, H. F. "Primary Affectional Patterns in Primates." *Amer. J. Ortho-Psychiat.* 30 (1960), 676–684.

———. "The Development of Affectional Patterns in Infant Monkeys." In B. M. Foss (ed.), *Determinants of Infant Behavior.* New York: Wiley, 1962, 75–97.

———. "Basic Social Capacity of Primates." In C. F. Southwick (ed.), *Primate Social Behavior.* New York: Van Nostrand, 1963, 153–161.

———. "The Primate Socialization Motives." *Trans. and Studies of the College of Physicians of Phila.* 33 (1966), 224–237.

IMANISHI, K. "Social Behavior in Japanese Monkeys, *Macaca fuscata.*" *Psychologia* 1 (1957), 47–54.

ITANI, J. "Paternal Care in the Wild Japanese Monkey, *Macaca fuscata fuscata.*" *Primates* 2 (1959), 61–93.

——— *et al.* "The Social Construction of Natural Troops of Japanese Monkeys in Takasakiyama." *Primates* 6 (1963), 1–42.

JAY, P. "Aspects of Maternal Behavior Among Langurs." *Ann. N. Y. Acad. Sci.* 102 (1962), 468–476.

———. "Field Studies." In A. Schrier, H. Harlow, and F. Stollnitz (eds.), *Behavior of Nonhuman Primates,* Vol. II. New York: Academic Press, 1965a, 525–591.

———. "The Common Langur of North India." In I. DeVore (ed.), *Primate Behavior: Field Studies of Monkeys and Apes.* New York: Holt, Rinehart and Winston, 1965b, 197–249.

JENSEN, G., R. BOBBITT, and B. GORDON. "The Development of Mutual Independence in Mother-Infant Pigtailed Monkeys, *Macaca nemestrina.*" In S. A. Altmann (ed.), *Social Communication Among Primates.* Chicago: University of Chicago Press, 1967, 43–55.

JOLLY, A. *Lemur Behavior.* Chicago: University of Chicago Press, 1966.

KAUFMAN, I. C. and L. ROSENBLUM. "The Waning of the Mother-Infant Bond in Two Species of Macaque." In B. Foss (ed.), *Determinants of Infant Behavior,* IV. London: Methuen, 1969, 41–59.

KAWAI, M. "On the Rank System in a Natural Group of Japanese Monkeys." *Primates* 162 (1958), 84–98.

KAWAMURA, S. "Matriarchal Social Ranks in the Minoo B Troop: A Study of the Rank System of Japanese Monkeys. *Primates* 2 (1958), 181–252.

KOFORD, C. B. "Rank of Mothers and Sons in Bands of Rhesus Monkeys." *Science* 141 (1963) 356–357.

KUMMER, H. *Social Organization of Hamadryas Baboons: A Field Study.* Chicago: University of Chicago Press, 1968.

———. "Dimensions of a Comparative Biology of Primate Groups." *Am. J. Phys. Anth.* 27 (1969), 357–366.

LOIZOS, C. "Play Behavior in Higher Primates: A Review." In D. Morris (ed.), *Primate Ethology.* Chicago: Aldine, 1967, 176–218.

LORENZ, K. "The Companion in the Bird's World." *Auk* 54 (1937), 245–273.

———. "Play and Vacuum Activities." In S. Auturi (ed.), *L'Instinct dans le Comportement des Animaux et de L'Homme.* Paris: Masson et cie, 1956.

MACROBERTS, M. "The Social Organization of Barbary Apes (*Macaca sylvana*) on Gibraltar." *Am. J. Phys. Anth.* 33 (1970), 83–101.

MASON, W. A. "The Effects of Social Restriction on the Behavior of Rhesus Monkeys: I. Free Social Behavior." *J. Comp. Physiol. Psychol.* 53 (1960), 582–589.

———. ". . . II. Tests of Gregariousness." *J. Comp. Physiol. Psychol.* 54 (1961a), 287–290.

———. ". . . III. Dominance Tests." *J. Comp. Physiol. Psychol.* 54 (1961b), 694–699.

———. "The Effects of Environmental Restriction on the Social Development of Rhesus Monkeys." In C. F. Southwick (ed.), *Primate Social Behavior.* New York: Van Nostrand, 1963, 161–174.

———. "The Social Development of Monkeys and Apes." In I. DeVore (ed.), *Primate Behavior: Field Studies of Monkeys and Apes.* New York: Holt, Rinehart and Winston, 1965, 514–544.

———. "Naturalistic and Experimental Investigations of the Social Behavior of Monkeys and Apes." In P. Jay (ed.), *Primates: Studies in Adaptation and Variability.* New York: Holt, Rinehart and Winston, 1968, 398–420.

MCNEIL, E. *Human Socialization.* Belmont, Calif.: Brooks-Cole, 1969.

MEAD, M. *Sex and Temperament in Three Primitive Societies.* New York: Morrow, 1935.

———. "Socialization and Enculturation." *Current Anthropology* 4 (1963), 184–188.

MENZEL, E. "Naturalistic and Experimental Research on Primates." *Human Development* 10 (1967), 170–186.

———. "Primate Naturalistic Research and Problems of Early Experience." *Develop. Psychobio.* 1 (1968), 175–184.

MITCHELL, G. "Paternalistic Behavior in Primates." *Psychol. Bull.* 71 (1969), 399–417.

MORRIS, D. "The Response of Animals to a Restricted Environment." *Symp. Zool. Soc. Lond.* 13 (1964), 99–118.

POIRIER, F. E. "The Nilgiri Langur (*Presbytis johnii*) Mother-Infant Dyad." *Primates* 9 (1968), 45–68.

———. "The Nilgiri Langur Troop: Its Composition, Structure, Function and Change." *Folia Primat.* 19 (1969), 20–47.

———. "Nilgiri Langur Ecology and Social Behavior." In L. A. Rosenblum (ed.), *Primate Behavior: Developments in Field and Laboratory Research.* New York: Academic Press, 1970, 251–383.

RIESEN, A. H. "Critical Stimulation and Optimum Periods." Paper read at A.P.A., New York, 1961.

ROSENBLATT, J., G. TURKEWITZ, and T. SCHNEIRLA. "Early Socialization in

Domestic Cats as Based on Feeding, and Other Relations Between Female and Young." In B. M. Foss (ed.), *Determinants of Infant Behavior.* London: Methuen, 1961, 51–74.

ROWELL, T. E. and Y. SPENCER-BOOTH. "Aunt-Infant Interaction in Captive Rhesus Monkeys." *Anim. Behav.* 12 (1964), 219–226.

SADE, D. S. "Some Aspects of Parent-Offspring and Sibling Relations in a Group of Rhesus Monkeys, with a Discussion of Grooming." *Amer. J. Phys. Anth.* 23 (1965), 1–17.

———. "Determinants of Dominance in a Group of Free-ranging Rhesus Monkeys." In S. A. Altmann (ed.), *Social Communication Among Primates.* Chicago: University of Chicago Press, 1967, 99–115.

SCHNEIRLA, T. and J. ROSENBLATT. "Critical Periods in the Development of Behavior." *Science* 139 (1963), 1110–1115.

SCOTT, J. P. "Group Formation Determined by Social Behavior." *Sociometry* 8 (1945), 42–52.

———. "The Social Behavior of Dogs and Wolves." *Ann. N.Y. Acad. Sci.* 55 (1950), 1009–1021.

———. "Critical Periods in the Development of Social Behavior in Puppies." *Psychosomatic Med.* 20 (1958), 42–54.

——— and M. MARSTON. "Critical Periods Affecting the Development of Normal and Maladjustive Social Behavior of Puppies." *J. Genet. Psychol.* 77 (1950), 25–60.

SHIMAHARA, N. "Enculturation—A Reconsideration." *Current Anthropology,* 11 (1970), 143–154.

SIMONDS, P. E. "The Bonnet Macaque in South India." In I. DeVore (ed.), *Primate Behavior: Field Studies of Monkeys and Apes.* New York: Holt, Rinehart and Winston, 1965, 175–197.

SOUTHWICK, C. F., M. A. BEG, and M. R. SIDDIQI. "Rhesus Monkeys in North India." In I. DeVore (ed.), *Primate Behavior: Field Studies of Monkeys and Apes.* New York: Holt, Rinehart and Winston, 1965, 111–160.

STRUHSAKER, T. T. "Auditory Communication Among Vervet Monkeys (*Cercopithecus aethiops*)." In S. A. Altmann (ed.), *Social Communication Among Primates.* Chicago: University of Chicago Press, 1967a, 281–325.

———. "Behavior of Vervet Monkeys (*Cercopithecus aethiops*)." *Univ. of Calif. Publ. Zool.* 82 (1967b), 1–74.

———. "Social Structure Among Vervet Monkeys (*Cercopithecus aethiops*)." *Behavior* 29 (1967c), 6–121.

———. "Behavior of Vervet Monkeys and Other Cercopithecines." *Science* 156 (1967d), 1197–1203.

———. "Correlates of Ecology and Social Organization Among African Cercopithecines." *Folia Primat.* 11 (1969), 80–119.

SUGIYAMA, Y. "Social Organization of Hanuman Langurs." In S. A. Altmann (ed.), *Social Communication Among Primates.* Chicago: University of Chicago Press, 1967, 221–237.

TANAKA, J. "Social Structure of Nilgiri Langurs." *Primates* 6 (1965), 107–122.

TINKLEPAUGH, O. "Social Behavior of Animals." In F. Moss (ed.), *Comparative Psychology.* Englewood Cliffs, N.J.: Prentice-Hall, 1948, 366–394.

——— and C. HARTMAN. "Behavior and Maternal Care of the Newborn

Monkey (*M. Mulatta, M. rhesus*)." *J. Genet. Psychol.* 40 (1932), 257–286.

WASHBURN, S. and I. DEVORE. "The Social Life of Baboons." *Sci. Am.* 204 (1961), 62–71.

WASHBURN, S. and D. HAMBURG. "The Implications of Primate Research." In I. DeVore (ed.), *Primate Behavior: Field Studies of Monkeys and Apes.* New York: Holt, Rinehart and Winston, 1965, 607–623.

Frances D. Burton

The Integration of Biology and Behavior in the Socialization of *Macaca Sylvana* of Gibraltar

Socialization operates on the total organism within a specific context. The processes of socialization are ongoing and may be termed "maturation within the social milieu." These processes can effect or develop the proper behaviors only in the measure that the organism is receptive to them. This receptivity is directly a function of a concatenation of personal events at a particular moment.

These events are simply summarized as biological development. The inherent qualities of protoplasm—irritability, reproduction, energy utilization, and so on—are specifically organized as the zygote becomes embryo becomes fetus within the limits of the genotype. The particular form this organization takes within the range of possibilities available is influenced by its exogenous and endogenous environments. These range from the state of the mother nutritionally, emotionally, and endocrinologically, and the nature of the maternal feedback system to the particular nature of the zygote's cells and the genic system that affects them. The more differentiated the tissue, the greater the organization of the systems, and thus the more susceptible the organism to a wider range of diversified influences. Susceptibility refers to sensitive periods—the moment(s) at which the matter, or target structure, has developed sufficiently to respond by growth, differentiation, and so on, to a given influence—exogenous or endogenous.

The neonate emerges with a repertoire of abilities and a range of potentialities as Coghill (1929), Carmichael (1928), Kuo (1967), and others have illustrated with their researches. For example, Carmichael (1928), in studying the fetal guinea pig, observed ontogenetic motor development similar to that which Coghill found in the salamander, and Kuo in the chick. The first movements occur as thrusts from the heart's beating. Body movements, as in the bird embryo, are myogenic and originate anteriorly and progress caudally. It is of great importance to note that these movements are myogenic and not neurally inspired. The myelinization of the nervous system permits increasing integration of movements, which in turn allows for new coordinations within the limits of the morphological potentials (monkeys cannot fly; dogs cannot grasp).

Seen from this, the epigenetic view (Kuo, 1967), the conjecture of fully developed *patterns* of behavior, which are inherited (Lorenz, e.g., 1965; Tinbergen, 1951) or located in the central nervous system (Delgado, 1965; Moyer, n.d.), present in the embryo and merely awaiting the proper releasing mechanism for their full manifestation, is superfluous and reminiscent of the ancient homunculus and other preformationist theories. As Schneirla (1965), Maier and Schneirla (1964), Marler (1966), Tavolga (1970), and especially Lehrman (1970) have recently shown, the old nature-nurture or exogenous-endogenous dichotomization of behavior is not intrinsic to animals. Behavior is the inventory of all that an animal does, and there is a unity of the organism within its context. Kuo (1967, pp. 84–85) has argued that this ancient dualism is an artifact of western European logic stemming from the Hegelian dialectic.

> Modern philosophical, social and biological thinkers of the Western World seem to have been unable to free themselves from the bondage of . . . thesis and antithesis: mind versus body, nature versus nurture, innateness versus learning, and so on. Since the time of Sherrington, Pavlov and Watson we have had: reflex versus habit (chain reflexes), reflex versus instinct, instinct versus habit, and conditioned versus unconditioned reflexes, and similar dichotomies.

The recrudescence of dichotomization is evident in recent primatology, particularly in the writings of Struhsaker (1968) and Gartlan and Brain (1968). Gartlan and Brain evoke nurture, or more modernly, environment, as the major influence in the behavior of a primate group, whereas Struhsaker argues that the genotype transcends any particular milieu for any given group of a species, so that behavior is consistent throughout. Although Aldrich-Blake (1970) has attempted to reconcile these views by attempting a synthesis of the extremes, the tendency is for ecologically oriented scholars to oppose ethnologically oriented ones. Perhaps our desire to formulate theoretical frameworks has outstripped our descriptive knowledge.

The purpose of this article is to describe and analyze the biological processes of socialization in *Macaca sylvana* of Gibraltar, with emphasis on the biological developments that precede, follow, and integrate with the social ones.

CONDITIONS OF THE STUDY

Two troops of *M. sylvana* range freely on Gibraltar. They are the charges of the Gibraltar regiment of the British army and are provisioned to approximately one-third of their nutritional needs. The provisioning is not only to supplement their natural diet but also to limit to some extent their foraging in the built-up areas. The natural habitat of these two troops includes rock faces, vegetational growth including shrub trees and bushes, and terraces made primarily of concrete. The Middle Hill (MH) troop may range from the summit of the rock at 1,350 feet to approximately 200 feet, while the Queen's Gate (QG) troop tend to restrict themselves to approximately 1,040 feet to 520 feet. Although free-ranging, both troops are tame; that is, they permit approach from humans, hand-feeding, and, in the case of the QG troop, will jump on people. Members of both troops do not permit humans to touch them, and will threaten and occasionally bite if such a gesture is made.

The presence of humans, whether tourists, residents, or military personnel, is a daily event. The extent of this contact is greater for the QG troop, which is maintained as a tourist attraction. The MH troop, which roams over a larger area, receives less contact, as they frequent inaccessible areas and remain in the forest growth, partially out of view, even when they are in their "tourist location." The monkeys are tended by an officer of the Gibraltar regiment and are the responsibility of this regiment. Effects of disease and injury are, therefore, attenuated. The number of animals is controlled, both in absolute size and sex ratio, surplus animals being sent to zoos. This factor no doubt accounts for the great longevity of the animals, mitigating against the natural predation of disease, injury, and fighting.

Observations were made almost daily, generally from 7:00 A.M., when the animals first appeared from their sleeping shelters in the rock face, until they returned to their sleeping places at 8:00 or 9:00 P.M., depending on the seasonal setting of the sun, with a break at some time during the forenoon. It was possible to accumulate 462 contact hours, 278 hours with the MH and 184 with the QG troop, during the period April 28 to July 30, 1970. Four people were engaged in observation, and the contact hours represent the total period of continuous observation irrespective of the number of field workers present at any one time. Observation distance ranged from 2 or 3 feet to 400 yards for which 8 × 40, 9.5° binoculars were employed. Still and motion pictures were taken and vocalizations recorded. Further observations in October 1970, and December 1970 through January 1971 were made and have been drawn upon for this report.

GROUP COMPOSITION [1]

At the end of the 1970 birth season, the troops together numbered thirty-three animals, both troops having a male to female adult ratio of 1 to 5.

[1] See Table 1, p. 58.

The animals would not tolerate more than one adult male, and in the past, when a male was in his fifth year, fighting broke out with the leader until solitarization or injuries or the threat of these forced the army to remove the loser (Holmes, p.c.[2]).

In the Gibraltar macaques, the onset of behavioral patterns was closely linked with age. Analysis is based on the natural division into age-grades, which are described below. The following are the categories for the animals:

Infant 1:	birth to three months
Infant 2:	three months to one year
Juvenile:	one year to three years
Subadult:	three years to four years (female); three years to five years (male)
Adult:	four years (female); five years (male)

INFANT 1: BIRTH TO THREE MONTHS

Neonate pelage is dark brown dorsally, and yellowish ventrally. Fingernails are almond shaped and black, but the skin of the feet, hands, face, perineum, and tail are light pink. The eyes are dark. The genital region is swollen at birth, undoubtedly as the result of maternal hormonal influence, but recedes after forty-eight hours. Birth blotches may be present on the face as dark red patches randomly positioned. The umbilicus, approximately one to two feet long, is left to dry out and break off. The ischial callosities are barely distinct from the perineal region; they appear as soft, wrinkled, pinkish translucent tissue within the smoother pink area. The tail is visible and approximately two to three centimeters long. As the individual matures the tail becomes proportionately smaller until in the adult it is not longer noticeable. The cheiridia are prehensile from birth, although the infant can only cling with its hands, which is probably why the mother supports the infant in the ventral clinging position as she moves about. Between the third and fourth day the feet have developed sufficiently for clinging but still appear weak, as the legs often dangle. Support is given less frequently, although the mother continues to signal the infant 1 that she is prepared to move by pressing its back. The infant 1 vocalizes from the first day of life—its first vocalization a high-pitched cry similar to the English sound "eeee." The infant 1 is capable of turning its head laterally, ventrally, and dorsally; but the latter two movements are uncontrolled, and the head wobbles as the movement is

[2] p.c.: personal communication.

accomplished. Movement to the teat is undirected, but sucking itself is coordinated.

The eyes seem to focus twenty-four hours after birth, although no test of this could be made. The impression was received by the manner in which the neonates oriented their heads, especially when seeking the teat; that is, they would look for it, sight upon it, and move their heads directly toward it, the dorsal neck muscles only permitting a strong wobble. The newborn infant in *M. sylvana,* as among most cercopithecoids, attracts the attention of the entire troop. There was a marked difference between the MH and QG troops in whether or not other females with or without infants in arms were permitted to approach. In QG troop, other females were permitted to sniffle and touch the neonate, although they were not allowed to take the infant 1 from its mother until it was approximately six to eight days old. In the MH troop other females were not usually permitted to come close until the infant 1 was over ten days old, and even then the other females were not allowed to take the infant 1 from its mother, although such attempts were made. Often a struggle ensued, with the mother pulling on the forepart of the infant 1 while the other female pulled on the hind parts, the mother usually threatening the other female at the same time, and the other female making the "teeth-chattering" grimace (Hooff, 1967; abbreviated in this article as "chatter") in return to the threat. This facial expression is an open-mouth gesture with lips and cheeks contracted, mandible opening and closing rapidly, teeth contacting teeth, and, in some animals, tongue flicking against teeth, producing a clicking sound. The head [3] male, however, was permitted to approach and even take the neonate from its mother on the first day (MH, four instances) and certainly by the third day (MH, one instance) and fourth day (QG, two instances). This behavior is extremely important to the development of basic social behaviors. The head male who is the putative biological father (Holmes, p.c.) sniffs, licks, caresses, pats, and holds the infant 1. While inspecting it, he looks into its face and chatters. The mother, who sits close by the head male, occasionally grooming him as he holds the infant 1, joins in chattering to it. Each time the infant 1 makes a sucking motion, both parents immediately chatter to it. By the third day at the earliest, and fifth generally, the infant 1 is capable of returning weak chatter movements when it is chattered to, but invariably just moves closer to the animal chattering to it. That is, it returns sucking motions with an added component of pulling back the lips and opening and closing the mouth. Thus a preexisting action movement, already

[3] This term seems preferable to "dominant," "control," or "leader," as these terms include functions that are not unique to this animal.

Figure 1. A two-day-old infant in its mother's lap.

Figure 2. The head male (MH) encourages the infant to walk toward him.

Figure 3. The subadult male retains the infant 1 while the leader male (left) and the mother (right) stay close-by.

included in the repertoire by birth, is conditioned into a purely social gesture. The importance of this lies in the use of the chatter, which functions as a distance-decreasing mechanism (Marler, 1968). It is (1) an approach gesture; (2) a pacifying and appeasement gesture; and (3) an invitation or encouragement to follow or be picked up. It is used when one animal is at a distance from another and moving in his direction. If a mild threat is given upon its approach, the animal will again chatter. If an animal surprises another by coming up from a place obscured from view, he will chatter. If two or more animals are wrestling in play, and one gives a high-pitched vocalization, the other(s) will chatter.

On as early as the sixth day of life, but no later than the eighth day, in the case of the MH troop, or fourth to eighth days, in the case of the QG troop, the infant begins to walk. It has already been climbing over its mother's and the head male's bodies since as early as the third day, mostly by means of its hands and arms. On four occasions the MH head male was the initiator of the infant's beginning to walk: he placed the animal on the ground, moving backward away from it to a distance of approximately two feet, lowering his head, and looking at the infant and chattering to it. The infant 1 would return the chatter and make crawling motions toward him. That is, each arm was alternately extended, the digits flexed, the legs and feet digits flexed, and a pulling-hopping motion employed. As the infant 1 approached within six inches to a foot, and if no other animal except the mother was nearby, the head male would again move away and repeat the chatter. If other animals began to close in, he would pick up the infant, and move away from the crowd, making a mild threat gesture to them.

It is interesting that encouraging the infant 1 to walk occurred after the infant recognized the chatter and was beginning to be able to use it, and before the infant's locomotor development permitted the action. Walking was initiated in the infant 1's first week by the head male in MH but not in QG, although the leader male in the latter troop did encourage infants to walk in the manner described after they had begun to do so on their own.

Significantly, at one and one-half months, the MH infants seemed capable of locomotor feats such as jumping, climbing, and leaping across two feet of space, with greater facility than did age-mates in QG. They differed in certain behavioral traits as well, particularly in greater apparent independence from their mothers and from male care. This was manifested by the fact that the MH infants could go farther away from surveillant animals and remain away longer than their QG age-mates. The age and personality differences of the leader males that fostered these differences are discussed below in the section on adult males.

From the fourteenth to the twentieth days after birth, the subadult males came to predominate as an influence in the infant's life. Three- and four-year-old males tended to follow the head male when he had the infant 1 from as early as the sixth day. After the infant had been walking for a few days, however, neither the head male nor the mother would interfere when a subadult attempted to pick up the infant 1. The first time this occurred, the head male and the mother would follow the subadult very closely and remain within a few feet of him. They relaxed their vigilance after a few days, especially after another female gave birth. First the leader male, then the mother would permit the subadult to take the infant as much as fifteen to twenty yards away. The distance increased to approximately 200 yards (during troop movements) when the infant reached a month old.

At first the infant 1 would give high-pitched vocalizations when the subadult took it. The subadult would chatter to it and eventually the infant 1 would leave the proximity of its mother or the head male in response to the subadult's chatter. Although it was during the time that the infant 1 was tended to by the head male that it learned to climb and be transported on the back, it was under the subadult's care that this behavior developed.

Like walking, the dorsal transport was initiated by the head male. This behavior began just after the infant 1 had begun regularly to walk toward the head male in response to the chatter. As the infant 1 approached him the leader male would turn sideways to the infant 1 and drop one shoulder so that the infant 1 in effect had no surface to touch other than the shoulder; the infant then climbed up. At first the infant 1 would lie horizontally across the back, or obliquely to the vertical axis, but within a few days it would consistently ride with its hands cephalad and its feet caudad. The infant 1's mother similarly would offer her shoulder, usually the day after the head male had just done so, but she tended to continue to carry the infant 1 ventrally (when she had possession of it) until after it reached approximately one and one-half to two months of age. As the infant grew larger the invitation for dorsal transport included a total body crouch or lowering of the hindquarters toward the infant 1 while chattering in its direction. At one and one-half to two months the infant 1 might solicit dorsal transport by touching the other animal's hindquarters or shoulder.

The subadult male stayed with the infant 1 as it made the maturational transition from adult surveillance to peer group membership—a period of approximately two and one-half to three months. As already mentioned, this association began once the infant could chatter and had developed basic locomotor skills, that is, walking, running, and climbing. The active role of the subadult in "male care" (Deag and Crook, 1970) was to "baby-sit"; that is, (1) to convey the infant 1 over any walking distance greater than approximately five feet, or over any jumping distance greater than one foot; (2) to remove the infant 1 from negative situations (physical danger, as in falling; threat of aggression, for example, from a female; or overstrenuous play of the infant 2); and (3) to provide the context within which the next stage of socialization takes place. The subadult seemed to convey the infant 1 always in respect first to the head male, then to its mother, irrespective of the distance from them. The subadult typically watched these individuals and if they should move into an area obscured from view, the subadult would often move in that direction until visual contact was restored. The subadult would often bring the infant 1 to the head male, chatter to the male, and with him to the infant 1. He then would either leave the infant with the head male or go off with it again. If the mother approached him, the subadult would either permit her to take the infant or would move off again with it. If the mother pursued, he would relinquish the infant to her.

The physical contact with the infant 1 was reduced as a function of the infant's growth: the older it became, the less the subadult would pick it up or offer his shoulder for dorsal transport; but the subadult's vigilance persisted until approximately six months (Holmes, p.c.). That is, as the one- to two-month-old infant clambered through the trees and over railings and rocks, or walked over uneven terrain, the subadult would be within one to three feet. The subadult would often follow along a branch, with one arm curved over and around the infant 1, his hand beneath it or his body over it, as if in readiness to grab it should it fall. The socialization of the infant 1 at this stage (fifteen days to three months) occurred largely within the framework of this relationship, but was not confined to it. The infant 1 would be with the subadult for a large part of the daylight hours, ranging from approximately three to four hours a day at one month to a peak of about eight hours at two and one-half months, this decreasing again at about three months to approximately four to five hours.

A subadult-infant association was not a fixed one. Like most of the relationships among *M. sylvana,* age, season, breeding status, temperament, and so on, affected the duration, intensity, and even the nature of the relationship. That is, as the infant 1 matured, it was passed from subadult male to subadult male. For example, Paris (three years old) took Rosemary when she was fifteen days old; when she was one and one-half months, Ben (four years old) took her, as Paris was occupied with Dorothy. By the time Olga was born, Eliot (three years old), the subadult male who had come from the QG troop, was no longer being threatened away from infant 1s, and while he accompanied Rosemary, Ben got Dorothy, and Paris was making the first approaches toward taking Olga from Mark.

During these months contact with the juveniles and infant 2s increased in proportion to the infant 1's age, from almost no contact at all at fifteen days, to about two hours per day broken into smaller time units at two months, to approximately four hours similarly broken down at three months. While the actual contact remained of short duration until about three months of age, the infant 1 was close to the juveniles, held or carried by the subadult or within a few inches to about five feet from him. The orientation toward the juveniles and infant 2s was reinforced by the rejecting behavior of the head male toward the infant 1. At anywhere from as late as one month of age to only, but not less than fifteen days, depending on the advent of new infants into the group, when the infant 1 would run up to the head male, or when the subadult surrogate would bring the infant 1 to the head male, instead of chattering to it, holding it, or carrying it, the head male immediately returned it to its mother, who then released it. The subadult, who had followed the head male, would then chatter to the infant 1 to follow or get onto its back. If the head male was holding a younger infant when an older infant approached, the head male might mildly threaten the older infant or the subadult who was bringing it toward him.

The increased association of the infant 1s (fifteen days to three months) with the juveniles and infant 2s coincided with the period of greatest development of locomotor and manipulative skills. This was also the time when the infant 1 was physically capable of imitating the younger animals. It was during this period that the infant began to test and, after one and one-half months, eat solid foods. Although many nonfood objects were brought to the mouth (rocks, dirt, paper), the infant 1 tended to put to his mouth those foods that he saw other animals take. Juveniles and infant 2s typically gathered foods (like flower heads, stems and, in the dry season, pulp of twigs) that are softer than those taken by subadults (like leaves and, in the dry season, tubers and roots dug up from six to ten inches in the ground). Although there was, of course, an overlap in the diets (such as seeds from various sources, figs, fruits, and so on), the infant 1 would try both the harder and softer foods, and was observed to reject, for example, thick leaves in favor of a diet closer to that taken by infant 2s and juveniles. No doubt this feeding behavior was related to masticatory muscle and tooth development, milk teeth being fully erupted by five months, first molars by ten to twelve months (Fitzgerald, 1950). Juvenile and infant 2 play with the infant 1s differed from that with age-mates in that the animal appeared to restrain himself, that is, he did not fully use his strength as he would in playing with members of his peer group. If the infant 1 should squeal during these bouts, the subadult would interfere, picking it up and/or giving a mild threat to or chasing away the juvenile or infant 2.

Gradually during its association with subadults and younger animals, the infant 1 would learn the appropriate responses for specific vocalizations in given contexts. This process seemed to be a conditioning one, a function of the differential behaviors of the subadult male, such as his picking or scooping up the infant in response to a negative vocalization (warning bark, threat, and so on) or his lack of a particular response to a positive vocalization (greeting call, contentment noise, play sounds, and so forth).

INFANT 2: THREE MONTHS TO ONE YEAR

At three months pelage has uniformly lightened with the fine lanugo hairs being gradually replaced by shorter ones. Eye color is beginning to lighten, and the infant 2 appears to have trebled in length from its size at birth.

All locomotor movements are coordinated by three months; the animal can walk without the lateral wobble characteristic of this motion at one and two months. Digits of cheiridia are extended instead of flexed in locomotion. Leaping and manipulation are, however, still clumsy, perhaps because depth vision essential for these has not yet developed, but the legs have gained muscular strength, so that even though a leap may miss its mark it is executed correctly. By approximately three months the male genitalia are distinguishable from the rear as a thick amorphous sac, yellowish-white in color.

In the Gibraltar groups, weaning in the nutritional sense began at about three months, with the mother responding more slowly to the infant's signals (for example, squealing, nuzzling) to be taken to the teat. Reorientation away from the head male and the mother began even earlier.

Encouraged away from its mother from about the tenth day by the head male, the infant 2 would increasingly be brought into contact with other members of the troop in a descending age pyramid, until finally by the third month, the infant 2 frequented a play group of peers. This integration into the troop was effected by the action of the older animals until the infant was about three months, when he actively sought contact with age-peers, older infant 2s, and juveniles. The infant 2s active contact included jumping, pulling, wrestling, and so forth, with the older animals. Bodily contact with the mother through this grade diminished from approximately twelve to fifteen hours per twenty-four-hour day at three months to approximately eight to ten hours per twenty-four-hour day at six months. Socialization, then, at this stage had three qualities: (1) it took place primarily within the context of the younger animals; (2) the animal was the recipient; and (3) the animal was the initiator. The study period did not cover the critical time between the third and eighth months. This developmental period—which covers the complete change from long dark pelage to the short, fuzzy, tawny gray of the older infant 2s, complete nutritional weaning, the refinement of locomotion and manipulation, as well as the refinement of response to vocalizations—must await future work and analysis. Some general statements, however, may be put forward.

The first two qualities of interaction, listed above, have been discussed in an earlier section. The active participation characterized the young infant 2. The tentative approaches to infant 1s, which had previously been rebuked by mild threats or retreat of the younger animals' mothers, were now permitted. The behaviors of touching, pulling, hitting, and climbing, which had been reinforced in the context of the subadults, juveniles, and older infants close to their first birthday, were transferred to this group of age-mates. The approach to peers may be attributed to "curiosity" however defined, to low intensity approach stimulus (Schneirla, 1959), or to contact resulting from random exploration. It is, of course, impossible to assess without experimentation what the young infant perceives of the subtlety of interrelationships in his society, or when a perception previously gained becomes a cognition or influence.

Peer contact at this stage was characterized by surveillance, primarily by subadults, head male and mother coequally, and juveniles, in that order. As the individuals of the infant 2 peer group developed in size and strength, distinguished primarily by the change in pelage, interference with the playing group diminished. That is, squeals and vocalizations would generally receive less rapid response than the cries of an animal one or two months younger. Booth (1962) has discussed the significance of the neonate pelage to the troop. It is, judging by older animals' response, a mark or signal of psychophysiological changes intimating, perhaps, the increasing ability of the infant 2 to take care of itself. This was evident when an infant of this age group would remove *itself* from a play situation that was too physically strenuous, leap across spaces that were two and one-half feet wide without assistance, climb and descend twenty-to-thirty-foot trees, and so forth.

Beyond the age of six months the faces deepen in color, but are not yet fully pigmented. The eyes are lighter than those of the infant 1. The ischial callosities have only just begun to differentiate into the adult patterns (see below). Cheiridia (hands and feet) are fully pigmented. The tail is visible as a stub one-half to one centimeter long. Pelage has lightened to a tawny gray, and the hair is quite short and gives the appearance of fuzziness. The face is round and there is yet no development in the superciliary region.

Sexes can be told apart from the rear, as the scrotum is no longer amorphous and gives the appearance of a sac with two chambers. The scrotum is large enough to be evident but not obvious. The penis is clearly visible ventrally, and flaccid; it seems to be approximately two to two and one-half centimeters long.

Vocalizations are still high pitched, though lower than those of the infant 1.

In the Gibraltar troops infant 1s were observed to make tentative grooming motions at one month, and by nine months the infant 2s groomed regularly. Although executed correctly, that is, with chatter gesture and hand, finger, and mouth movements described for Cercopithecidae in general, the activity was initiated less frequently than by other age groups. Each grooming bout was sustained for a very short period, never in excess of five minutes (see Table 2, p. 59). With four males and one female in this age group, it is not possible to say whether or not there was any sex difference in the amount of time spent grooming. Infant 2s groomed and were groomed most by adult females, but received grooming more from subadult males than from the head male, although they groomed the head male more frequently than they did the subadult males.

Bouts of grooming with peers, juveniles, and younger infants easily turned into play sessions, with either groomer or groomed initiating. An infant 2 groomer involved with an older animal (subadult or adult) was easily drawn away from this activity by his peers or juveniles. Infant 2s preferred to play with peers or juveniles rather than with subadults, who often initiated or joined these play groups. "Preferred" refers to both the relative duration of a play bout and the fact that when a subadult joined younger play groups, the younger animals would move away from the subadult and continue group play activity or do something else (individual play, groom, feed, and so on). The significance of this is that the tendency to frequent peer groups was maximal in infant 2s and juveniles and, largely because of sex differences in behavior, diminished thereafter until, by the adult stage, there were no discrete groups, only individuals.

The infant 2 was still occasionally transported dorsally by the subadult males. Either the subadult or the infant 2 might initiate contact for this—the subadult looking behind and perhaps lowering his hindquarters, and the infant 2 by grasping the subadult from the rear at about the level of his pelvis. When the subadult did not lower his hindquarters, the infant 2's head would only reach to just above the subadult's back, so that the infant 2 was obliged to climb up; he did this generally by placing a foot about the level of the popliteal cavity, so that the infant 2's penis was at about the level of the subadult's callosities. As the infant 2 completed the movement onto the subadult's back he might thrust his pelvis as in copulation. Whether or not he did this because in climbing upon the subadult he was stimulated by contact is difficult to assess, as copulatory mounting arises at approximately the same age in *Cercopithecus* that do not generally carry dorsally. Nevertheless, in *Macaca sylvana,* the proper mount seems to develop from a precursor movement, just as the chatter develops from precursor sucking motions. In this context it is significant that the subadult female who does not carry infant 1s is solicited for dorsal transport by the infant 2s, and this transport is often accompanied by pelvic thrusting.

In summary, the *critical* biological development in the infant 2 age group is the change of pelage that inaugurates qualitatively different interactions, characterized by increasing independence of the infant. Sexual dimorphism has not yet appeared and behavior is basically homogeneous throughout the peer group, with the important exception that the one female spent more time sitting with, or grooming, the adult females, especially her mother.

JUVENILE: ONE YEAR TO THREE YEARS

The juvenile group is a transitional one. In many respects, the juvenile could be considered an "infant 3" on the basis of size, degree of sexual dimorphism, and play-group mates. In terms of participation in surveillance behavior, sexual behavior, and types of association with older animals, however, they have certain affinities with subadults that warrant the designation "juvenile."

The juvenile, by his second birthday, appears approximately a third longer than the first year old. This seems to be a behaviorally critical time, but in external development there is little morphological contrast to the first year old. The fur is beginning to change to the adult condition; that is, it is somewhat longer and gives less appearance of fuzziness. The light pinkish facial coloration has given way to brown. Freckles, white spots, and so on, which distinguish one adult animal from another, have begun to appear. Eye color is dark hazel to light brown. The ischial callosities have differentiated, the male pattern being an ovoid continuous shape, whereas the female pattern is one of two distinct, nearly triangular forms, whose apex is central. The difference is easily noted and enables rapid identification. The scrotum is visible from the rear, but extends only approximately one inch toward the knee. The female perineal region appears as a circular mass distinct from the callosities region. In the groups studied there was as yet no sex difference in size, but some differentiation in behavior was manifest that perhaps reflected sex hormone development. Vocalizations had deepened, and the "eeee" fear sound was replaced by "ahnh" with the exception of one male juvenile who retained the infant sound because it would bring the head male or subadult to or near him. This lasted until two months before his birthday.[4]

The juvenile male would mount subadult males and females and give pelvic thrusts with greater frequency than would the infant 2, but the juvenile male still elicited dorsal transport from the head male or from the subadult males (see Table 3, p. 60). That is, while this behavior was transforming into sexual behavior, it retained the quality of its infantile function. The juvenile female (QG troop) did not mount other animals. While all juveniles groomed more frequently than did infant 2s (see Table 2), the females sustained the bouts at least three to five minutes longer but received less grooming and as a corollary were decreasingly involved in play. When not involved in grooming or play, the juvenile female might be sitting close to an adult female, or feeding and sitting by herself. The juvenile female tended to begin the solitary existence that characterized subadult females (see below). She spent much of the day sitting or wandering by herself, whereas the male was more integrated and more likely to be groomed, playing, or simply associating with others. The juvenile female was more likely to be repulsed from contacting infant 1s or 2s, and infant 2s were more likely to avoid play invitations. The juvenile could displace the infant 2 from food (both natural and provisioned), play, place, and grooming, but it would be more exposed to censure from the subadults and full adults. This punishment took the form of being threatened, chased, grabbed, pulled, or a combination of these, with the juvenile responding by a distress vocalization, appeasement chatter, submissive crouch, or moving off. The distress vocalizations might bring another subadult or adult to the juvenile, and this older animal might intercede with the punisher (threatening, and so forth) or might make contact with the juvenile (embracing, chattering, grooming).

Although each animal is given a name at birth by the British army, unless the animal is with its mother it cannot be told apart from age-mates until three years of age or later, because the freckles, pelage patterns, behavior characteristics, and so forth, that distinguish one from another are not sufficiently developed until that time. It was therefore impossible to tell whether any intercession by an older animal was effected because of a

[4] Similar behavior had been observed in *C. aethiops* and *C. neglectus* at Tigoni Primate Research Centre, Kenya, where an animal caged with its biological parents would retain the vocalization of a younger animal. Responses of older animals were protective toward it. When, in each case, the animal was moved to a different cage with age-mates, the vocalizations appropriate to its age group were then utilized. It is as if the animal in question were manipulating a behavioral response. Lack of response by the age-mates seemed to have extinguished this behavior.

particular genetic, familial, or other special bond. (See MacRoberts, 1970, for a different view.) It is perhaps more likely that a mother would intercede for her offspring, and a subadult male intercede for the infant he had carried for a month, but this does not clarify the head male's intercessions (all the offspring were his) nor the subadult female's intercession (she very seldom carried infants) or clarify *which* animal the subadult would intercede for, as the two older subadults both carried two infants during the course of this study. Intercession was an important element of socialization, as it established the boundaries for permissible behavior in given contexts and conditioned the animal to appropriate behaviors vis-à-vis younger animals. Furthermore it was instructive as a means by which the juvenile came to learn the personalities of the older animals. The juvenile began to relate to the older animals in terms of their idiosyncrasies rather than in terms of functional categories such as "male-care," protection, and so on. For example, the juvenile had to learn that this female (Joan, the youngest, MH troop) was more apt to grab if an infant 2 squealed than was another female (Bridget), or that one subadult male (Paris) was more apt to chase than was another (Ben), who would only mildly threaten.

Juveniles participated in troop life on a different level. For example, if a dog, hawk, or other threat appeared, the juvenile might give the warning bark. Interestingly, the response to a juvenile warning signal was observed to be investigation of the situation rather than immediate flight or defense posture. That is, the adults and subadults would look in the direction the juvenile was looking, and then respond. As discussed from the point of view of the infant, the juvenile was involved with the care and socialization of younger animals. At first threatened away by the mother, leader male, or subadult in the male-care role, the juvenile was increasingly allowed to contact the infant as it matured. When the infant was approximately three months old, the juvenile might offer dorsal transport, engage the infant in play, and so on. Similar behaviors extended from the juvenile to the infant 2.

The juveniles never initiated troop movements. The direction having been established (see below), the juveniles would walk with or follow the subadults. They might also walk along with an infant 2, who in turn might be accompanying a female, walking with a subadult male, or walking in a peer group. As the subadult males were more directly involved in troop movement, the routes must have been memorized sometime between the late infant 2 stage or during the juvenile stage. The routes taken did not vary; there were two down to the Moorish Castle area and back to Caroline's Battery, and two up to Middle Hill, with use depending on season.

SUBADULT MALE: THREE TO FIVE YEARS OLD

A substantial change is apparent between an animal beginning its second year (older juvenile) and one beginning its third year (subadult). The latter is twice the size of the second year old, and this is particularly visible in the limbs, which have quite nearly attained their adult length. The face has lengthened, particularly in the nasal area, more so than in the older juveniles. The facial characteristics are apparent; spots, blotches, and pigmented areas, which permit recognition of individuals, are now present. The fur is quite long, coarse and dry, and more interspersed with brown. Ischial callosities are fully differentiated, and the scrotum extends from approximately one inch to almost two inches toward the knee. Males have a characteristic walk, an undulating, almost salamander-like motion. The body is still slim, full bulk not being achieved till beyond the seventh year (Holmes, p.c.). Vocalizations are quite deep and the repertoire has expanded to include the true warning bark—a deep-throated staccato "rowr" akin to a dog's bark in rhythm and sound.

The influence of sex hormones on the behavior of the subadult male may be deduced from the difference in the nature of its sexual experimentation as compared to those of the juvenile or infant 1. Direct stimulation of genitalia was sought from others, the self, or objects. Mounting occurred with increased frequency, and was distinct from dorsal transport in the posture assumed, which was close to the fully adult one. Occasionally, however, both feet grasped the distal portion of the femur, or both feet might be on the pelvis, or one foot up and one down. Subadult males mounted subadult females occasionally, and other subadult males and the leader male frequently, but they were not observed to mount adult females probably because this would lead to a fight with the leader male (Holmes, p.c.). Animals of both sexes, from juvenile to infant 1, were held in the correct posture and thrust against, although a correct mount could not occur because of the younger animal's size. Two such episodes were observed to result in ejaculation after the younger animal was released. Masturbation, while less frequent in the summer months than during the breeding season of September to December (Holmes, p.c.), did occur, and ranged from rubbing the genitalia against a stone to autogrooming an erect penis, often culminating in ejaculation. Grooming, particularly of the ventral and femoral areas, frequently caused an erection, the penis appearing to be approximately two inches long.

Play was less acrobatic and included more wrestling and chasing, occupying fewer of the daylight hours than did grooming or sexual behavior as compared with the younger animals. The two instances recorded of subadult male play with the leader male are significant because neither was more than two minutes long, and both took the form of wrestling and bumping, with the leader male crouching low to the ground with a chatter face. In these instances, the subadult males were restrained in their movements. That is, muscular tension and jerky movements characterized the interaction that in each case terminated in a chatter embrace. This controlled play is in contrast to bouts between peers or with younger animals. Subadult males were never observed to play with either subadult or adult females. While total play time decreased at this age, grooming increased. Subadult males groomed adult females more frequently than they did subadult females, but groomed the leader male almost twice as frequently as both female groups combined. The subadult male could approach neonates when the leader male held them, and attempts to gain proximity to them by means of grooming him contributed to the high frequency.

Subadult male contact with the subadult females changed radically during the course of the study and progressed from virtually total disregard of them to infrequent bouts of grooming, and more infrequent mountings. Toward the end of the study period, however, subadult females were permitted to stay near the subadult males when they had, or were close to, the infant 1s. Interactions with the adult females also changed. While the subadult males could still be displaced by the adult females at the end of the study, the frequency decreased, and the behavior was more often met with resistance in the form of mild threat. Furthermore, subadults gave way to certain adult females and displaced certain others. For example, they were more likely to stand their ground against Joan, the youngest female, than against Caroline, a ten year old. Subadults would intercede for infants and juveniles when these were chased by adult females—by threatening or chasing the female—and look at or for the head male as if to see whether he would intervene on behalf of the females. The adult male did not respond to this behavior toward the females.

Aggressive incidents among subadult males themselves were slightly more frequent than those among younger animals. These incidents took the form of wrestling and chasing, as did play, but were accompanied by threat gestures and vocalizations. The subadult males scrapped primarily over the taking or keeping of an infant 1 or 2, which contact began after the infant could stand alone and had been chattered to and encouraged to walk. Thus this kind of contact could begin as early as the seventh to the twelfth day of the infant's life. Only two of the subadult males were permitted to care for infants. The third and youngest male (Eliot) had come from the QG troop. His exclusion from contact with the young infants may reflect the fact that, although he first appeared at MH in February 1970 (Holmes, p.c.), he was not yet a fully integrated member of the troop; when the infants were two and one-half to three months old, he was permitted the male-care role.

When the infant 1s were less than a month old, the subadults would return to the head male with them, give and receive a chatter or chatter-embrace, and then depart again. After that age, the head male would often threaten the subadults away. This behavior seemed to have a double function: (1) it was effective in reorienting the infant toward younger animals; and (2) it acted to reinforce the subadults' leadership behavior. That is, the subadults' responsibility for the infant grew the more he was rebuffed. The fact that at times the subadults were found more than two miles from the troop also attested to their growing independence.

The subadult males took an active role in troop movement. Like the adult females and head male, they could indicate time and direction of movement by moving in the desired direction and sitting on a high point with the body facing the direction. Whether or not they then "led" the movement, they usually moved in the vanguard. If the head male led the movement, they might be a considerable distance (up to 200 yards) from the infant's mother, though close to the head male. The subadult males further assumed troop responsibility by "standing guard," that is, when people, dogs, and cats approached or a warning bark was given indicating something that might not be visible, the subadult males would sit near or facing the danger. If a female threatened some such intruder, the subadult males, themselves threatening, would rush toward the thing being threatened.

To summarize, the subadult male's socialization at this stage was less a learning of new skills or practicing of old ones than a transformation of those already learned with the addition of increased independence and responsibility. He retained an ambivalent position vis-à-vis the adults, that is, like a juvenile he approached them for chatter-embrace, to be groomed, to groom, and so forth. While difficult to objectify, the subadult's body posture reflected this ambivalence. Many approaches to adults were submissive, with all muscles tense and the body in a slight crouch. At the same time, however, the subadult male began to assert himself toward the adults, by, among other things, threatening, particularly when intervening on behalf of a younger animal. His assertions would be met with more and more deference, which undoubtedly reinforced this behavior. The leader male was influential in this acquisition of independence and leadership by not interfering when the subadults were threatening females, and by rebuffing them when they returned infant 1s to him.

SUBADULT FEMALE: THREE TO FOUR AND ONE-HALF YEARS OLD [5]

The females are generally smaller than the males of the same age; this difference does not occur before this stage. Their faces, too, are smaller and rounder, though similarly marked with blotches, freckles, and so forth. The ischial callosities are by now two distinct triangular forms, and the perineal region, which is pigmented, has enlarged to a diameter of approximately five centimeters; there is no other sign of sexual development. In other physical respects they very much resemble the subadult males. The Gibraltar subadult females walked with a straight-through movement and very little undulation of pelvic or pectoral girdles. The interraction of subadult females with the other animals was characterized by timidity and infrequency. These females were typically solitary, wandering or eating by themselves. Their primary contact with other animals was in grooming, although they were never observed to touch each other or the head male. Unlike the subadult males the females could be displaced by all adults and subadult males.

Subadult females played with the infant 1s and 2s, but with no other group. Play consisted primarily of wrestling and chasing and was generally solicited by the subadult female.

The contact pattern between the subadult female and the infant 2s or juveniles was perhaps significant in the subadult female's socialization, as in many respects it simulated mother-infant relationships. The subadult females solicited contact from the infants in play and grooming. They attempted to hold them, although the young ones would break away; and they offered dorsal transport, which was sometimes accepted as such, but which frequently became a sexual mount. Subadult females were attracted, as were all other troop members, to the neonates, and would even try to grab them. They were consistently rebuffed by adult females, the head male, and subadult males when they approached the infant 1s. Toward the close of the study, however, they were allowed greater proximity to the neonates and were even permitted to touch them for moments at a time. Whether this represents a recognition of the maturation of the subadult female or personality factors in the adult females who had babies at this time can only be known from subsequent study.

Sexual behavior began to change as the females approached their fourth birthday. Previously mounted only by infant 1s and 2s, as they approached their birthdays the subadult males began to mount them. Although the

[5] Four and one-half years is given as the upper limit for this age group, as females are considered to be adult after parturition, which has occurred as early as four years of age.

action was accompanied by pelvic thrusts, there was no intromission. The adult male was not observed to take any sexual interest in females of this age group during the first study period, which terminated before the breeding season. Unlike the subadult males, the females were never seen to mount each other or to engage in any form of autoeroticism. Both females had their first sexual swelling by October, 1970, which differed from those of the adult females in color and form. Two distinct swellings were visible: one around the anus, and one apparently involving the labia. The skin of the uppermost swelling appeared taut while that of the lower area was looser. Unlike that in the adult females, the color was gray, white, and pink. Adult female swellings involve the entire perineal area and are larger and blacker. These swellings were present in December, 1970, and in the case of the slightly older female, resulted in an infant born at the beginning of June, 1971. This infant died on its fifth day, apparently due to her mother's insufficient milk production.

Subadult females, like adult females, gave the warning bark, often after an older animal had done so. They did not lead troop movements and usually took a place in the middle of the migration, often following an adult female, or accompanied by a younger animal.

It is very difficult to understand what was involved in the social processes of subadult female maturation. Since subadult females were solitary for the greater part of the day, learning would not seem to be a function of association. Perhaps the hormonal changes occurring at this time are in themselves causative of changes in behavior, or are influential in conjunction with the subadult female's perception and subsequent identification of herself as a female. In order to avoid anthropomorphism, this kind of question should be studied in the laboratory rather than under field conditions. What is known from analyzing the British army records, and from the instance herein reported, is that subadult females can conceive at as early as three and one-half years. Inasmuch as the social status of a female adult is largely a function of having an infant, such a subadult female would be socially recognized as an adult by just over four years of age.

In summary, subadult females appeared to be isolated from the rest of the group even though they groomed and were groomed as much as some other age groups, because they did not seek each other's company, and when their invitations to play were accepted, the play was shorter in duration than was play on the part of the subadult males. They were almost totally ignored by the leader male, except when they threatened or disturbed one of the young. The isolation seems to be part of the overall limitation of the female's role in the socialization of the young, as elaborated in the next section on adult females. As the subadult female becomes sexually viable, the frequency of contact with males increases—first with the subadult males and then, in full estrus, assumedly with the leader male. But it is probably having an infant that fully integrates the female into the troop, as a function of troop interest in the neonate.

ADULT FEMALE: FOUR AND ONE-HALF YEARS AND ABOVE

Adult animals are included in this discussion both because they illustrate the end point of socialization and because biological development, even though catabolic, does not cease with maturity; behavior changes throughout the life span of the animal according to its previous experiences.

Adult females are characterized by bulk, by distinctive markings about the face and head, and by the perineal prominence. The facial markings include individualistic pigmentation of the face, particularly around the eyes, where freckles and nonpigmented spots may be variably distributed. The superciliary area is especially well marked, as a function of the increase in bone and fur. The perineum swells in the breeding and birth seasons, and may become so large that the tightened epidermis cracks or tears (QG troop), leaving large gaping areas with no epidermal covering. Outside of these seasons, the perineum remains prominent and is usually darkly pigmented except for the longitudinal section lining the vaginal and anal orifices. The fur at the sacrocaudal juncture forms an arcuate pattern as a result of the swelling, and is useful in identifying a seated female at a distance. After parity the teats remain distended, sometimes to a length of approximately an inch. They are variably pigmented, with black, white, and pink areas forming bands or spots. During lactation the breast tissue is quite distended, primarily below the level of the teat. The distance between them is approximately three to three and one-half inches. The infant suckles from one teat at a time.

Adult females of the groups studied, perhaps as a function of their bulk, did not leap from tree to tree as did the younger animals. They could execute acrobatic feats, and were observed to make jumps of ten to fifteen feet. They seldom did this, however, and tended instead to lower themselves down from heights to cross the ground, particularly if carrying an infant.

For females taken as a group, the period of solitarization terminates with maturity. Adult females were not observed to engage in any form of play, to the extent that they would (unlike *Cercopithecus* females) reject overtures or inclusion in play with threats or cuffs. During the summer study period the adult females spent approximately 85 percent of fourteen daylight hours grooming or being groomed (see Table 2). Typically, grooming bouts were not only more frequent, but of longer duration than those of other age groups. Association was not, however, restricted to grooming, as females would sit, feed, and sleep with other troop members. This generalization does not hold for all the females, however, as personality factors seemed to affect the nature and amount of interaction with other individuals. Wilma and Joan of the MH troop and Fiona of the QG troop, for example, tended to be more solitary throughout the study than did the other females. Caroline and Bridget (MH) tended to frequent each other's company, as did Venus and Rose (QG). Charlotte (MH) could move freely from female to female even when that female had just given birth. The factors involved in the differences in conduct are hard to define, although "personality" or "temperament" identify the nature of the factor. Wilma, for example, was very submissive, to the extent that she would retreat from contact offers. Joan was very energetic, and would rush or move quite quickly toward other animals, which would often result in her being threatened or in the other animal's moving away.

Adult females participated in troop protection, but the amount of their participation seemed to depend largely on group composition. That is, in the MH troop whose head male was fifteen years of age, and which contained two subadult males and one more in the process of integrating into the troop, the function of troop protection appeared relegated, for the most part, to the males, although the females had some share. In the QG troop, however, the leader male of which was only five years of age, and which was losing its only subadult male, the adult females adopted the role of guardian in standing watch, warning of danger, interceding for the young, and so forth.

Troop movements were primarily initiated by the head male or adult females. An adult female would stand or sit in an open area or on a high point where she was clearly visible, or she might begin walking slowly in a particular direction. Some subadults and/or infants would come next; the leader male would then follow. When he went off, the remainder of the troop would move on in the chosen direction. Even when not leading, however, the adult females exercised strong control in troop movement. If one did not follow the troop, the adult male might call to her or return to the female in question and then lead off again. If she still did not follow, he might return, and the troop would follow him back. There was no indication of how the troop discriminated between the male's returning to urge the female to follow, and his leading the troop back. Adult females were observed to aggress toward all members of the troop (including infant 1s) with the single exception of the head male. They were the subject of aggression primarily from each other, and secondarily from the adult male. The most frequent causes of strife were attempts at contacting the infant 1s and proximity to the head male. Most aggression took the form of chasing and vocalizing, although aggravated situations would turn into true fights, characterized by grabbing, pulling, and biting, unless the leader male intervened. Most of the aggression from the head male was in the form of intervention, the male threatening or chasing a female who was aggressing toward some other troop member. Occasionally his intervention developed into grabbing at the female, or even biting. Generally a look and mild threat from him, or even his presence alone, was sufficient to reduce tension.

As stated earlier, data from British army records indicate that females may be impregnated for the first time at three and one-half years, although sometimes they will not necessarily conceive or successfully carry until six years of age. Females observed to be pregnant during the study have subsequently been noticed by the Officer-in-Charge to be no longer carrying. As the fetus will have been aborted in the bushes, verification is inadequate. During the summer of the 1970 study, one female (QG) lost her fetus from prolapsed uterus (M. Sutton, p.c.), and another, who had been pregnant early in the study, appeared no longer to be carrying by the middle of the summer study period (Holmes, p.c.).

It is assumed that all females are impregnated only by the head male (Holmes, p.c.), as the mounting of adult females in the breeding season by the subadult males incites aggression from the head male. Copulations occur throughout the year, but probably are not "true" copulations (Carpenter, 1942), because none of the copulations observed (thirty-five in MH, nine in QG) was seen to culminate in ejaculation. Episodes involving ejaculation may have occurred, however, outside the area of observation. In three instances, one in MH and two in QG, copulations took place just before, on the day of, or just after a birth, and seemed perhaps to function more as reinforcement of ties or "reassurance" than as reproductive sex.

The female, as described in the section on infant 1s and 2s, has a limited role in socialization of the young. The female might, however, act as a negative force, as, for example, did Bridget (MH), and Deirdre and Tessa (QG), who retarded contact with their respective head males by removing their infants a distance from him, so that encouragement of chattering, walking, climbing, and so on, took place later than with the other neonates. It seemed that these infants' development in locomotor and manipulative skills particularly took place more slowly. Dorothy (Bridget's infant), for example, was not any more capable at locomotor tasks (jumping distances, and so on) at one and one-half months than was Olga (Charlotte's infant), then aged only two weeks. Nevertheless, the female serves primarily to reinforce learning. That is, consistently the day after the male has encouraged the infant 1 to walk, or to climb dorsally, the female does the same. There were no observations of any female doing these sorts of activities before the male had done so. As the female would relinquish the infant first to the male and then for greater periods of time to the subadult males, her contact with her own newborn was severely reduced after the sixth day or so. The infant is returned to its mother when it squeals a great deal, perhaps because it is then hungry. Assumedly, the infant sleeps with its mother until it is past six months, although it was impossible to observe this directly. Her function in the troop then, other than the obvious reproductive one, was reflected by the grooming frequencies, and would seem to be primarily as social adhesive. The male was the focus of the troop, but it was through the female that the infant was imprinted to its troop, since the female's body is the first environment of the infant, and since from the first day of its life the female sniffs, licks, grooms, and generally touches it. Perhaps because of its biological need of her as a young animal, and through the grooming association thereafter, it is through her that this connection with the troop is predominantly reinforced. (See Koford, 1963; Sade, 1965.)

ADULT MALE: FIVE YEARS AND ABOVE

A male is biologically an adult at five years of age, as he can reproduce at that time, although there is one report (Fitzgerald, 1950) of a male (Mick) fathering offspring at three and one-half years. The gradual process of developing bulk begins slowly after age three and accelerates within the fifth year. The canines, which have erupted by two years (Fitzgerald, 1950), descend below the occlusal plane, and the scrotum becomes larger and more pendulous, extending approximately three and one-half inches toward the knee. The superciliary region develops heavily after the fifth year, so that the fur may dangle forward. Spots and other characteristic marks are fully developed on the face, to which may be added fight scars. The fur is slightly darker than in the adult female, coarse, dry, and very long; it is quite full around the head and shoulders, appearing like a mantle. Between the fifth and seventh years, greater bulk is achieved, and this increase in girth continues to nine years. An adult male may eventually weigh approximately sixty to seventy pounds (Holmes, p.c.), and his canines may reach a length of approximately one inch from the gum. The adult male has lost the swagger, the undulating walk of the subadult male, and moves with a deliberate even motion. Like the female, and perhaps for the same reason, the adult male seldom leaps; he is cautious in climbing, and has been observed to test a branch or twig before placing his weight on it.

The attainment of leadership in Gibraltar is largely manipulated by the Officer-in-Charge. That is, when a young male approaches five years of age, a decision is made as to whether the incumbent leader can and should go on, or should be removed to a zoo in favor of the younger animal. The present leader of the QG troop was taken from the MH troop and put in the position at just under five years to fill a vacuum, as the former head male had died and the next possible male, then about thirteen years of age, had previously joined the MH group. This older male at first shared the troop with the incumbent, then eleven years old (Holmes, p.c.). After the thirteen year old (Mark) had been caged so that a hand injury could be tended, the two males began to fight, and the fighting became so serious that the younger one (Harold) was removed and put in a cage. Mark was thus chosen as leader of the MH troop, the choice having been made in his favor largely because Harold had once killed an infant, and on another occasion so molested a female that she thrust her infant away from her. The present QG leader (Sam), now fully five years old, has begun to fill his role. At times, however, he joins the subadults and infants in games, and the adult females have been observed to assume and share the leader role for the period. The head male is responsible for the protection of the troop as a whole and particularly of the younger infants. Although he does not necessarily initiate troop movements, he controls them by following (or not following) the initiator. The manner of leading troop movements has been described under the section on adult females.

The adult male functions as the "control" animal, as described by Bernstein (1966, 1968). In Gibraltar the adult male restrained individuals from aggression or broke up fights that had gotten beyond the threatening stage. He spent most of the day observing the surroundings and surveying the troop, this even while grooming or being groomed. He was apparently alert, even when dozing, as the slightest noise would cause him to sit up or move toward the noise to investigate. He played an important role in the socialization process as described in the section on infant 1s. The nature and "definition" of the role was apparently a function of the head male and his age. An adult male may live until seventeen or twenty years of age (Holmes, p.c.), and much of what he does would seem to be the result of experiences gained. Mark (MH), for example, aged fifteen years, differed greatly from Sam, the five-year-old leader of QG. For example, Mark seldom played with the younger animals, whereas Sam frequently did. Mark seldom did more in punishment of an infant than threaten, and Sam would chase, grab, or bite. Both males immediately became alert when an animal made a distress call, but Mark waited longer to move in that direction, at first just turning his head toward the sound and looking. As he responded to some situations and not others, it was as if he were evaluating the nature of the situation. The first and second year olds in Sam's group approached a newborn infant within the first four days of its life. Mark would threaten off such approaches, until the infant was at least five to six days old. Although both males conditioned the infant to chatter, and both males took the infant within the first few days of its life, Mark would encourage the infant to walk before it had the motor coordination to do so. In distinct contrast to this, in Sam's group the infants began to walk largely on their own initiative, undoubtedly largely because their mothers often removed them from contact with Sam. Interestingly enough, infants of one and one-half months in Mark's group seemed to have better locomotor coordination than did corresponding infants in Sam's; that is, they moved forward with less lateral wobble and could climb and jump without falling over.

If "respect" is operationally defined as speed of reaction to another individual combined with intensity of stimulus necessary to elicit this reaction, Sam received less respect from his females than did Mark, as Sam had to engage in physical contact or combat instead of mildly threatening them before they would respond, move, and so on.

In summary, the adult male's role in socialization is: (1) to encourage the infant to develop motor abilities that permit social interaction; (2) to reorient the infant 1, as it matures, away from himself and toward other troop members; (3) to reinforce socially acceptable behaviors appropriate to the age group by not interfering, or by giving positive reward (chatter, embrace, and so on); and (4) to extinguish or negate inappropriate behaviors by punishment (threat, chase, and so forth). Furthermore, in his functions as leader male of the troop the adult male acts as (1) major or sole breeder, (2) control animal, and (3) focus of the troop.

DISCUSSION

The major features of each age-grade's role in socialization are summarized herein. The adult female is the first environment for the neonate, and thus the context for the most important biological maturation. The infant first sees and makes its first oriented movements within her arms. However, since the leader male takes the neonate from as early as the first day of life, it is the leader male who is the preeminent influence in socialization of the infant until it is approximately two weeks of age. The leader male encourages biological maturation by reinforcing the infant's mouth sucking movements until they become the social chatter gesture, and by encouraging the infant in locomotor skills and to take the dorsal transport position. These basic body movements are of prime importance for all future social contact. The leader male reorients the infant away from its mother, himself, and other adults, by permitting subadult males to first snatch the infant, and later to take it away for greater and greater distances. It is while the infant is under the care of the subadult males that it increasingly contacts juveniles and later age-mates. From contact with subadult males and juveniles, as well as from this social context, the infant learns, as he develops motor skills, all the necessary information to become a functioning member of the troop. The information ranges from what to eat to available routes from one area to another. Subadult females play a very small role in socialization of the infants, as this age group is largely isolated from other troop members, and as they are rebuffed by all adults and subadult males when they make overtures. The adult females are more important in socializing juveniles and subadults than in socializing infants. Their relationship to the infants is primarily as feeder, and later groomer, that is, a relationship of comfort and support. Although they do defend and protect the infants, this is more the task of subadult males. Females influence the older animals primarily by rebuffing them as they approach infants, play with juveniles too roughly, or come for food the female herself is approaching. But what they learn from these females seems more to be how to respond to the personalities and temperaments of the individual animals.

What has been described is a population as it appeared at a particular point in time. It is interesting that notes on the Gibraltar monkeys made in the late 1940s indicate that certain biological and social features were different. At the time the notes were made, there were two strains of monkeys and their mixed descendants: one, smaller in number, descended from macaques on Gibraltar prior to World War II; the other, descended from African imports made to secure the population. A major difference between the 1940 monkeys and the present ones is relevant to this discussion. Infant socialization was apparently not as limited to the male context, as females, particularly sibling females, were at that time permitted to touch and hold each other's infants (Fitzgerald, 1950). The change in tradition and transformation in role definition implicit in this fact serves to stress the point of the plasticity of social relations even within a circumscribed deme (or population). In other terms, while the infant needs contact for "normal" development, as deprivation studies have shown (Harlow, 1962), there is no prescription as to which individual must occupy the contacting role. The personality of the head male is a large factor in this, particularly when he remains the leader for a prolonged period and thereby has access to several generations. In a sense then, there may be a "tradition drift" (Burton and Bick, in press), processually akin to genetic drift. The QG troop may serve as an example. As Sam gains bulk, completes his canine development, and gains experience, assuredly the adult females in his group will respond to him differently. The new infants maturing at that time will therefore be exposed to relationships substantially different from those described above. It may be that he will, for example, take infants from their mothers when they are younger, and thus the acceleration in motor development seen in MH troop may become a feature in QG. The females' responsibility for the troop may be circumscribed as Sam takes on more responsibility himself, such as standing guard duty more frequently, and so on.

The difference just described illustrates the fact that there is flux in a population. This article is intended to delineate the extent to which biology, environment, tradition, and social relations alter and interaffect each other, giving rise to new forms. To see behavior as genetically fixed, or wholly environmentally determined, would be to ignore the basis of evolution.

TABLES

The following tables are included to represent tendencies in, but not the entirety of, the activities quantified. In most cases such overall representation is impossible, even over extended contact periods as was the case in Gibraltar. The animals are highly mobile, and the duration of an activity cannot always be ascertained or calculated, because although the activity may begin in direct view, it may continue or terminate behind a bush or in a tree, where visibility is, at best, limited. For example, infant 2s and juveniles spend most of the day, probably 90 percent, playing. Yet a frequency chart for play of this age group would not suggest this, and it is because when these young animals play, they range over the total area the troop is in, much of it bush. The factor of selection must also be admitted. Even with three observers posted at different places in order to maximize observation and obtain a check on the data, some kinds of actions will be noted and not others—the most unusual behavior, for instance, will be recorded and the most ordinary neglected, as reflected in the relatively low frequency of mother-infant grooming. This problem, which existed under positive observation conditions, is probably exacerbated by poor viewing conditions as with forest monkeys. Frequency charts are extracted from the context in which the activity occurred, and most activities are tightly interrelated with others. In one block of time, five or more activities may go on, interrelated or giving rise to each other, but each activity would be entered on a different chart, with the effect that the quantification itself does not adequately convey the texture of that which is analyzed.

TABLE 1. Group Composition: April 28 to September 23, 1970

	INFANT 1S		INFANT 2S		JUVENILES		SUBADULTS		ADULTS		TOTAL
	Male	*Female*	*Male*	*Female*	*Male*	*Female*	*Male*	*Female*	*Male*	*Female*	
Middle Hill	0	4	4	1	2	0	2(1*)	2	1	5	22
Queen's Gate	0	2	0	0	1	1	1(0*)	1	1	5	11
Total	0	6	4	1	3	1	3	3	2	10	33

* Subadult male moved from Queen's Gate to Middle Hill during the study period.

TABLE 2. Grooming

GROOMER	GROOMED							TOTAL
				Subadult		*Adult*		
	Infant 1	Infant 2	Juvenile	Male	Female	Male	Female	
Infant 1								
Infant 2			1 *.06%*	14.5 *.91%*	2 *.13%*	19 *.19%*	32.5 *2.04%*	69 *4.34%*
Juvenile			18 *1.13*	18 *1.13*	1 *.06*	23 *1.44*	72.5 *4.55*	132.4 *8.31*
Subadult Male		28 *1.76%*	15.5 *.97*	11 *.69*	5 *.31*	134 *8.41*	83 *6.59*	350.8 *22.01*
Subadult Female		9 *.56*	29 *1.82*	1 *.06*			145.5 *9.13*	184.5 *11.57*
Adult Male		15 *.94*	1 *.06*				279 *17.50*	274 *17.19*
Adult Female	9 *.56%*	140.5 *8.81*	25 (13/.82*) *1.57*	39 *2.45*	80.5 *5.05*	90.5 *5.68*	205.5 *12.89*	578 *36.26*
TOTAL	9 *.56*	202.5 *12.70*	102.5 *6.43*	83.4 *5.23*	89.5 *5.61*	330.4 *20.73*	783.1 *49.13*	1,594

NOTE: Arabic numbers equal grooming time in minutes; italicized arabic numbers equal percentage of total grooming time of groomed animal.
* Sex of adult unknown.

TABLE 3. Development of Sexual Motions

	FREQUENCY OF DORSAL (TRANSPORT) MOUNT WITH THRUST				FREQUENCY OF PELVIC MOUNT			
	Subadult		*Adult*		*Subadult*		*Adult*	
	Male	Female	Male	Female	Male	Female	Male	Female
Infant 1	0	0	0	0	0	0	0	0
Infant 2	8	0	0	0	4	2	0	2
Juvenile	15	0	0	0	7	1	0	0
Subadult Male	0	0	0	0	11	2	11	0

Acknowledgments

The research described in this article was made possible by a Humanities and Social Science Research Grant from the University of Toronto.

I would like to extend many thanks to Sergeant Alfred Holmes of the Gibraltar Regiment, Officer-in-Charge of monkeys, for tutoring me in the way of life of the monkeys and for generally assisting me with them. His extensive knowledge and comprehension of them was invaluable.

The British army was most generous in permitting me to undertake and in facilitating this research. Their kindness in extending special permission and in providing equipment and access to the documents on the monkeys made this research possible. I am most grateful to Brigadier General Michael Wingate-Gray, Major Dederick Wright, Major Domingo Collado, and especially Major Nicholas Carter.

I would also like to thank the Royal Air Force for enabling me to examine meteorological data.

Three people were with me during most of the research period. I owe a great debt to Richard Doble, Bill Gibson, and particularly Patricia Tiberius for their precise observations, frequent discussions, meaningful criticisms, and astute insights. Their achievements are recognized with profound gratitude.

A very special acknowledgment is extended to Suzanne Rozell, whose patience, editing skills, and thoughtful suggestions were of very great assistance.

Mr. Ben Sousan, curator of the Gibraltar Museum, helped identify local flora and fauna and was most considerate in availing me of documents. Mrs. Barbara MacSporran and Cathy Lowinger very kindly typed the manuscript.

References

ALDRICH-BLAKE, F. P. G. "Problems of Social Structure in Forest Monkeys" in J. H. Crook (ed.), *Social Behavior in Birds and Mammals.* London: Academic Press, 1970, 136–158.

BERNSTEIN, I. S. "Analysis of a Key Role in a Capuchin (*Cebus albifrons*) Group." *Tulane Studies in Zoology* 13 (1966), 49–54.

———. "Some Observations in a Wild Troop of *Macaca irus.*" *Folia Primatologica* 8 (1968), 121–131.

BOOTH, C. "Some Observations on Behavior of Cercopithecus Monkeys" in J. Buettner-Janusch (ed.), *The Relatives of Man.* New York: New York Academy of Science, 1962, 477–487.

BURTON, F. and M. J. A. BICK. "A Deme in Time." *J. of Human Evolution* (in press).

CARMICHAEL, L. "The Development of Behavior in Vertebrates Experimentally Removed from Influence of External Stimulation." *Psychol. Rev.* 33 (1928), 51–58.

CARPENTER, R. "Sexual Behavior in Free-ranging Rhesus Monkeys, *Macaca mulatta.*" *J. Comp. Psych.* 33 (1942), 113–162.

COGHILL, G. E. "The Early Development of Behavior in *Amblystoma* and in Man." *Arch. Neurol. Psychiat.* 21 (1929), 989–1009.

DEAG, J. and J. CROOK. "Social Behavior and 'Agonistic Buffering' in the Wild Barbary Ape, *Macaca sylvana*" (in press).

DELGADO, J. M. R. "Sequential Behavior Induced Repeatedly by Stimulation of the Red Nucleus in Free Monkeys." *Science* 18 (1965), 1361–1363.

FITZGERALD, CAPT. Personal notes on the apes (1950).

GARTLAN, J. S. and C. K. BRAIN. "Ecology and Social Variability in *C. aethiops* and *C. mitis*" in P. Jay (ed.), *Primates: Studies in Adaptation and Variability.* New York: Holt, Rinehart and Winston, 1968, 253–292.

HARLOW, H. F. "The Effect of Rearing Conditions on Behavior." *Bull. Menninger Clinic* 26 (1962), 213–224.

HOOFF, J.A.R.A.M. VAN. "The Facial Displays of the Catarrhine Monkeys and Apes" in D. Morris (ed.), *Primate Ethology.* Chicago: Aldine, 1967, 7–68.

KOFORD, C. B. "Rank of Mothers and Sons in Bands of Rhesus Monkeys." *Science* 141 (1963), 356–357.

KUO, Z. Y. The Dynamics of Behavior Development. New York: Random House, 1967.

LEHRMAN, D. S. "Semantic and Conceptual Issues in the Nature-Nurture Problem" in L. Aronson *et al.* (eds.), *Development and Evolution of Behavior.* San Francisco: Freeman, 1970, 17–52.

LORENZ, K. Z. *Evolution and Modification of Behavior.* Chicago: University of Chicago Press, 1965.

MACROBERTS, M. H. "The Social Organization of Barbary Apes (*Macaca sylvana*) on Gibraltar." *American J. Physical Anthropology* 33 (1970), 83–100.

MAIER, N. R. F. and T. C. SCHNEIRLA. *Principles of Animal Psychology.* New York: Dover Press, 1964.

MARLER, P. *Mechanisms of Animal Behavior.* New York: Wiley, 1966.

———. "Aggregation and Dispersal: Two Functions in Primate Communication" in P. Jay (ed.), *Primates: Studies in Adaptation and Variability.* New York: Holt, Rinehart and Winston, 1968, 420–438.

MOYER, K. E. "Kinds of Aggression and Their Physiological Basis." Pittsburgh: Carnegie–Mellon University, Department of Psychology Report No. 67–12.

SADE, D. S. "Some Aspects of Parent-Offspring and Sibling Relations in a Group of Rhesus Monkeys, with a Discussion of Grooming." *American J. Physical Anthropology* 23 (1965), 1–17.

SCHNEIRLA, T. C. "An Evolutionary and Developmental Theory of Biphasic Processes Underlying Approach and Withdrawal" in M. Jones (ed.), *Current Theory and Research on Motivation.* Lincoln: University of Nebraska Press 7 (1959), 1–42.

———. "Aspects of Stimulation and Organizing in Approach/Withdrawal Processes Underlying Vertebrate Behavioral Development" in D. S. Lehrman *et al.* (eds.), *Advances in the Study of Behavior,* vol. 1. New York: Academic Press, 1965, 1–74.

STRUHSAKER, T. T. "Correlates of Ecology and Social Organization Among African Cercopithecines." Paper prepared in advance for participation in Symposium No. 42, Social Organization and Subsistence in Primate Societies, August 19–28. Wenner-Gren Foundation for Anthropological Research. 1968 Summer Season. (1969).

TAVOLGA, W. N. "Levels of Interaction in Animal Communication" in L. R. Aronson *et al.* (eds.), *Development and Evolution of Behavior.* San Francisco: Freeman, 1970, 281–302.

TINBERGEN, N. *The Study of Instinct.* Oxford, Eng.: Clarendon Press, 1951.

Neil R. Chalmers

Comparative Aspects of Early Infant Development in Some Captive Cercopithecines

INTRODUCTION

Primate mothers vary considerably from species to species in the type of care they give their infants. Among the Old World monkeys, perhaps one of the most striking differences to be seen is that between certain members of the Colobinae and some of the Cercopithecinae. In *Colobus guereza* and *Presbytis entellus,* on the one hand, Wooldridge (1969) and Jay (1963) have described how the newborn infant is handled freely by many members of the group, whereas in cercopithecines such as *Papio* and *Erythrocebus,* DeVore (1963) and Hall (1965) have shown how the mother is at first rather possessive toward her infant, and allows other members of the group at most only limited contact with it. Outside the Cercopithecidae, patterns of infant care can vary still more widely. For example, Fitzgerald (1935) and Moynihan (1964) have shown that for the New World monkeys, *Callithrix* and *Aotus* respectively, adult males may take a prominent part in caring for the infant at a very early stage. Such intensive care by the adult male is not commonly found in the Old World monkeys. Mothers may care intimately for their infants for many years, as in *Pan* (van Lawick-Goodall, 1968), or for only a few weeks, as in *Galago* (Sauer and Sauer, 1963).

Obviously, much of this variability is related to the phylogenetic position of the different species within the order, the mother-infant relationship tending to become more complex and longer lasting in higher primates. It is also possible, however, that other factors such as the type of habitat in which a species lives and the type of social organization that it displays affect the nature of the mother-infant relationship. These possibilities can best be investigated by focusing on a group of closely related primates, and seeing whether variations in these factors are accompanied by variations in mother-infant behavior.

In this article this procedure is adopted for five members of the Cercopithecidae, these being three members of the genus *Cercopithecus,* the closely related *Erythrocebus,* and the more distantly related *Cercocebus.* The work described here was carried out on captive animals. Although conditions of captivity are grossly different from those in the wild, this does not necessarily invalidate such work, for it is the differences between species that concern us here. If a variety of species kept under similar conditions of captivity show differences in behavior, then we may have some confidence that these differences are valid ones.

The article is divided into two parts. In the first part various aspects of mother-infant behavior over the first three months of the infant's life are described. These descriptions attempt to relate the differences in mother-infant relationships to differences in habitat. In the second part of the article some experiments are described that deal with the effects of social companions on mother-infant behavior. In this case, interspecific differences in such effects are explained in terms of differences in social organization and behavior, which themselves are of course related to differences in habitat.

COMPARATIVE DEVELOPMENT OVER THE FIRST THREE MONTHS OF LIFE

Infants of four species were watched over the first three months of their lives. These were the vervet (*Cercopithecus aethiops*), a predominantly terrestrial animal (Struhsaker, 1967b; Gartlan and Brain, 1968), De Brazza's monkey (*C. neglectus*), a species which, according to Simpson who is at present carrying out a field study of this species in Kenya, lives in riverine forest, and which frequently descends to the ground, particularly when frightened (p.c. [1]), Sykes monkey (*C. mitis,* also known as *C. albogularis*), a forest-living animal, frequently found in isolated forest patches, which Gartlan and Brain (1968) have described as predominantly arboreal, and the highly arboreal gray-cheeked mangabey (*Cercocebus albigena*) (Chalmers, 1968). Five vervet mother-infant pairs were observed, four De Brazza's, two Sykes, and four mangabeys.

All animals lived in groups consisting of one adult male, the mother and her infant, and one to four other adult females plus their young. They were watched in outdoor cages, the smallest of which (that for the Sykes) measured 6 × 3 × 2.5 meters, and the largest (that for the vervets) 9 × 3 × 2.5 meters. Details of the mother-infant pairs used are summarized in Table 1. All the adult monkeys had been living in captivity for at least ten months prior to these observations, and the majority for considerably longer. The animals were watched for one hour per day, usually between 11:30 A.M. and 12:30 P.M., three, four, or five days a week. Watches were not carried out during rain, for as has been shown (Rowell *et al.,* 1968), rain greatly modified the mother-infant behavior. The mean

[1] p.c.: personal communication.

TABLE 1. Details of Mothers and Infants Used in Infant Development Study

	MOTHER	INFANT	SEX	BORN	PARITY OF MOTHER
Vervets	L	Arthur	M	3/29/67	Multiparous
	S	Charlotte	F	4/20/67	Multiparous
	S	Eddy	M	1/6/68	Multiparous
	NT	Betty	F	5/31/67	Multiparous
	NF	Davy	M	10/5/67	Multiparous
Sykes	White	Zoe	F	11/27/67	Multiparous
	Orange	Y	M	6/13/68	Unknown
Mangabeys	Sophie	Lulu	F	9/19/68	Mother's 2nd infant
	Sophie	Coke	M	5/25/70	Mother's 3rd infant
	Stumpy	Mike	M	11/13/68	Multiparous
	Pet	Nobby	M	6/3/70	Primiparous
De Brazza's	Biddy	Bom	M	1/6/70	Unknown
	Bunty	Willy Boy	M	5/4/70	Unknown
	Alice	Ruth	F	2/24/70	Multiparous
	Tatty's Daughter	Oliver	M	5/1/70	Unknown

number of observation hours on the vervets over the three months was 50.6, for the De Brazza's 40.5, for the Sykes 44.5, and for the mangabeys 56.3. Observations were recorded on check sheets. These divided the hour's observation into half-minute periods, and for each half-minute, features of the mother-infant behavior were recorded such as the infant's position relative to the mother, whether the infant was awake or asleep, whether the infant was being groomed by the mother or not, and so on. Details of these recording methods are similar to and derived from those described by Hinde *et al.* (1964). Two-thirds of all the observations were carried out by the author, the remainder by trained assistants. Differences in recording techniques between the different observers were negligible.

Records for each individual monkey were summarized weekly, thus giving measures such as the amount of time per week the infant was asleep, the amount of time the mother groomed the infant, and so on. Figures were expressed either as percentages of time watched, or where more appropriate, as percentages of the infant's time awake. Such figures were used for interspecific comparisons, interspecific differences being tested for significance by Mann-Whitney U tests with a 5 percent level of significance. Because of the small numbers of Sykes involved, such tests could never reach significance when made between Sykes and mangabeys, or Sykes and De Brazza's (Siegel, 1956). Accordingly, Sykes were tested only against vervets in this way.

The observations on the vervets, Sykes, and two of the mangabey infants

Figure 1. Mean Percentage of Time Asleep per Week During First Thirteen Weeks of Life.

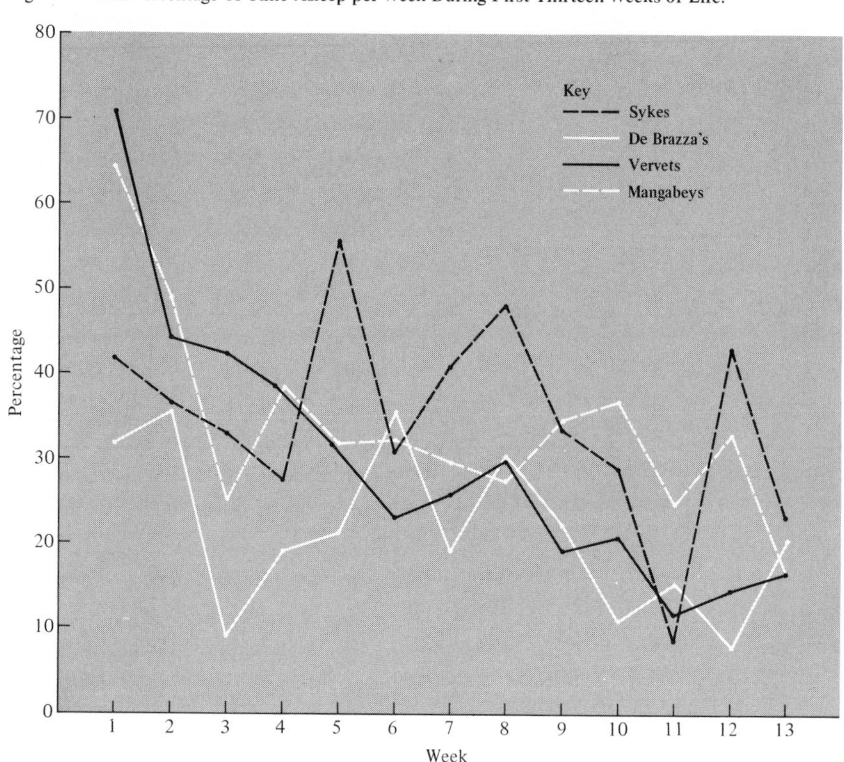

were carried out in Kampala, Uganda, where the average daily temperature was approximately 75° F., and where there was little seasonal fluctuation in temperature or rainfall. Observations on the other two mangabey babies and on the De Brazza's babies were carried out at Limuru, Kenya, where the average daily temperature was 65° F., and where seasonal fluctuations in rain and temperature occurred. The possible effect of these climatic differences on the results is discussed later.

RESULTS

Figure 1 shows the percentage of half-minute periods for which the monkeys were watched that the infants were asleep. The figures given in this and subsequent graphs are the weekly means of individual scores for each species. In most cases the proportion of time spent asleep does not conspicuously differ between the four species. De Brazza's tended to spend less time asleep than the other species, the figures for De Brazza's being significantly lower than that for vervets in the third and fourth weeks, and

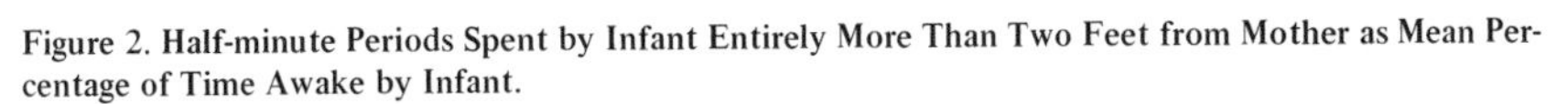

Figure 2. Half-minute Periods Spent by Infant Entirely More Than Two Feet from Mother as Mean Percentage of Time Awake by Infant.

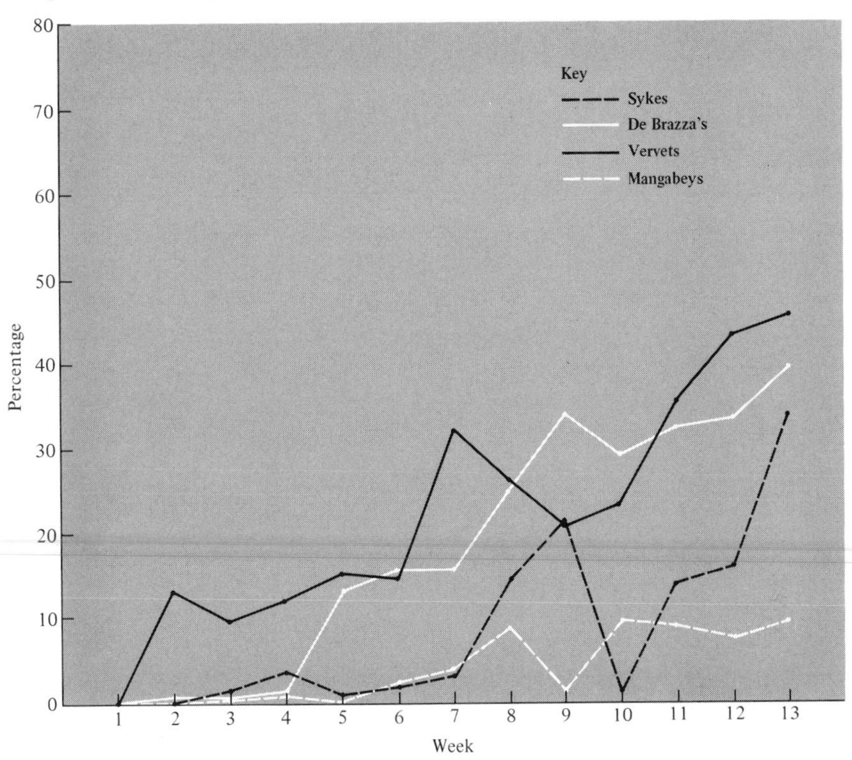

for mangabeys in the twelfth week, but no other differences are significant. The figures are also very similar to those for rhesus (Hinde *et al.*, 1964).

Although the amount of time the infants spent awake at a given age was for the most part the same, the behavior of the infants of the four species while awake varied strikingly. Figure 2 shows the number of half-minute periods during which the infants stayed more than two feet away from the mother, as a percentage of the number in which they were awake. The figure of two feet is chosen here as it represents the approximate distance within which the mother could reach out and touch the infant, and if necessary pick it up, without having to get up and walk toward it. The figures for vervets are very much higher than those for the mangabeys, the differences being significant from the second week onward, except for the eighth and tenth weeks. The figures for the vervets are also higher than those for the De Brazza's, but only significantly so in the third and fourth weeks. Figures for the De Brazza's are significantly higher than those for mangabeys in the ninth, tenth, twelfth, and thirteenth weeks. The figures for Sykes appear to be intermediate between those of the De

Figure 3. Number of Half-minute Periods in Which Mother Held Infant as Mean Percentage of Time Watched.

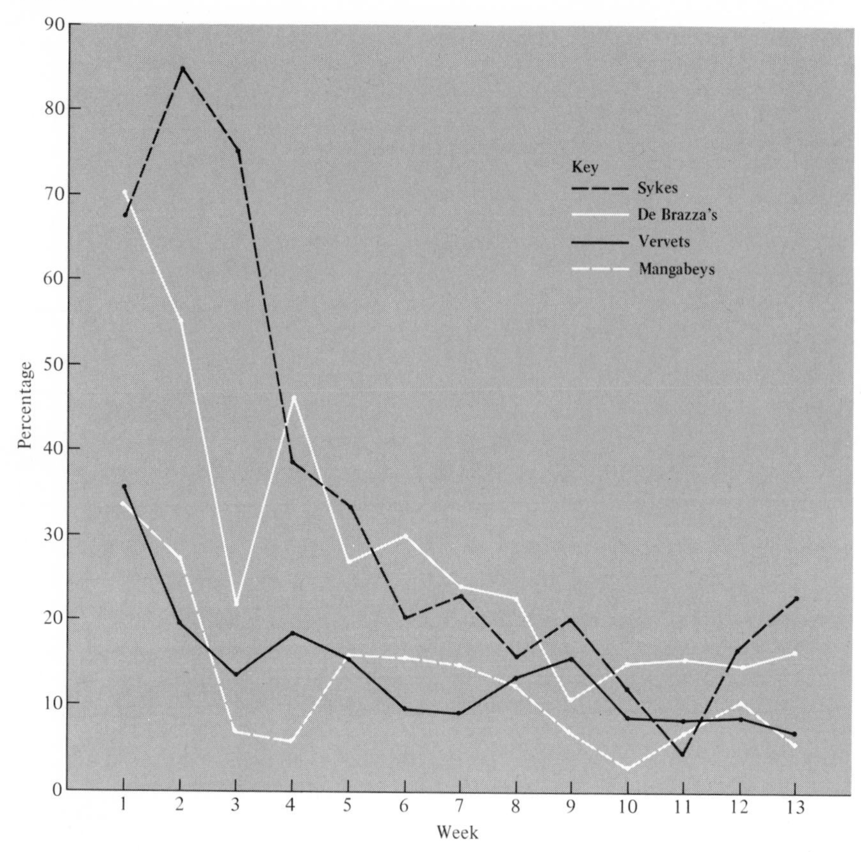

Brazza's and the mangabeys. The figures for the Sykes are significantly lower than those for the vervets in the twelfth week.

It seems that the more arboreal the species, the smaller the amount of time a baby of a given age spends at a distance from its mother. These differences between the species might be the result of one or both of the following possibilities:

1. Differences in the degree of restrictiveness of the mothers, the vervet mothers being the least restrictive, the mangabey mothers the most.
2. Differences in the behavior of the infants, vervet infants being the most adventurous, mangabey infants the least.

Figure 3 shows the percentage of half-minute periods that the mothers held their infants. When holding the infants ventrally, the mothers usually put one or both arms round their infants' shoulders, while the infant clung to them. In such a position the mothers could effectively prevent the infant from leaving. Figure 3 shows that mangabey and vervet mothers are very similar to each other in the proportion of time that they spend holding their infants. There are no significant differences between the two species. This suggests that the greater time that vervet infants spend away from their mothers as compared with mangabey infants resulted from the greater independence of the vervet infants from their mothers rather than the greater restrictiveness of the mangabey mothers. The figures for the De Brazza's are somewhat higher than those for vervets, but only significantly so in the fourth and seventh weeks. The De Brazza's infants in their fourth week spent a smaller proportion of their waking time at a distance from their mothers than did vervets, and this lower figure may have been caused by the mothers holding the infants more. Comparing "mother-holds" figures for De Brazza's and mangabeys, we find that those for De Brazza's are on the whole *higher* than those for the mangabeys, significantly so in the fourth and tenth weeks. The greater time spent off the mother by De Brazza's infants is thus achieved in spite of their mothers' restrictiveness. The figures for holding by Sykes mothers, while in later weeks are of the same magnitude as the other species, in the second and third weeks are considerably higher than the others. This was the result of the behavior of one of the mothers only. It may be that this contributed to the low figure for the percentage of half-minute periods totally off the mother in Sykes during the early weeks of life. However, the low figures for Sykes infants totally off the mother continue beyond the first five weeks of the infants' lives, right through to the end of the first three months, a period when the restrictiveness of the Sykes mothers is of the same order as the other mothers.

The differences between the species in the amount of time spent at a distance from the mother cannot, then, be explained solely in terms of differences in maternal restrictiveness, and must be due to a large extent to differences between the infants. One possible difference could be the maturation rate of the four species, vervet infants maturing the most rapidly, mangabeys the least. To find out whether this is so, the ages at which different criteria of motor activity were achieved are compared for the four species in Table 2. There is a difficulty here, however, for certain activities such as running, chasing, and wrestling might only be expected to be seen in infants that could move freely away from their mother. A mangabey or Sykes infant of a given age might well have the full equipment for these activities and yet not show them because it is for some reason inhibited from leaving its mother. It is probably wiser, therefore,

TABLE 2. Age in Days at Which Various Motor Activities First Seen

	VERVETS					MANGABEYS				DE BRAZZA'S				SYKES	
	A	*B*	*C*	*D*	*E*	*L*	*M*	*N*	*Ck*	*R*	*Bm*	*WB*	*O*	*Z*	*Y*
Steady locomotion	47	31	49	27	26	35	51	30	N.D.	26	18	38	42	34	34
Mean	36.0					38.7				31.0				34.0*K*	
Climbing	8	13	29	15	18	12	36	30	10	21	17	5	28	26	14
Mean	20.8					22.0				17.8				20.0	
Tries to leave	N.D.	6	N.D.	1	3	5	1	N.D.	10	N.D.	7	5	3	1	6
Mean	3.3					5.3				5.0				3.5	
Comes off mother	9	6	N.D.	1	3	2	1	N.D.	4	N.D.	8	2	4	1	6
Mean	4.8					2.3				4.7				3.5	
Picks up object in hand	8	20	44	9	26	34	36	10	11	N.D.	17	10	11	34	29
Mean	21.4					22.8				12.7				31.5	

N.D.: no reliable data.

to select criteria that do not substantially alter the infant's position relative to its mother. These categories of behavior need some explanation.

CATEGORIES OF BEHAVIOR

Steady Locomotion. In their earliest attempts at locomotion, infants tended to drag themselves forward (or push themselves backward) with the chest and abdomen close to the ground and the elbows and knees out to the side. As the infants got older their limbs straightened out and they were able to move confidently, without slipping or tottering. When this stage was reached it was classed as "steady locomotion."

Climbing. This always first occurred with the mother sitting against the netting of the cage, and the infant pulling itself up a few inches toward the mother's head along the wire netting.

Tries to Leave. In its earliest appearance, this usually took the form of the infant releasing its grip on its mother, twisting under her arm, and trying to move along her side and away from her. The mother usually prevented this by squeezing the infant firmly to her side with her elbow and/or holding on to its tail.

Comes off Mother. The infant sat by the mother, but maintained contact with her.

First Successful Picking up of an Object. This was usually achieved by a clumsy grip, involving all the fingers gripping against the palm of the hand, and did not involve the animal visually focusing on the object.

Table 2 shows that there is no evidence that these categories of behavior appear earlier in any one species than the others. That is, although vervet infants leave their mothers earlier in life than the other species, De Brazza's earlier than mangabeys or Sykes, and Sykes earlier than mangabeys, this is not because they vary in their physical abilities to leave the mother.

DISCUSSION

These results show that the infants of the four species have a comparable rate of development, and also spend a comparable proportion of time awake at given ages up to three months. However, vervet infants move away from their mothers to a much greater extent than do mangabey and Sykes infants. The extent to which they move away appears to be of the same order as baboons (Rowell *et al.,* 1968) and rhesus (Hinde *et al.,* 1964). De Brazza's infants still move away from their mothers more frequently than do Sykes or mangabey infants. It is clear, then, that the differences between vervets and De Brazza's, on the one hand, and Sykes and mangabey infants, on the other, are due primarily to a reluctance of Sykes and mangabey infants to leave their mothers rather than an increased restrictiveness of their mothers.

As mentioned earlier, the climate in the two areas where the study was carried out differed. It is possible that the colder weather that occurred when the De Brazza's and two of the mangabeys were watched might have increased the time the infants spent clinging to their mothers because of their need for warmth, thus invalidating interspecific comparisons. This is unlikely, however, because the two figures for the cold weather mangabey infants did not differ markedly from those for the two warm weather ones. Moreover, if the cold weather had any effect on the De Brazza's, the infants achieved a higher degree of independence from their mothers than Sykes or mangabey infants in spite of it.

In that vervets and De Brazza's spend much of their time on or near the ground, while the other two species are arboreal, Sykes considerably so and mangabeys almost totally so, it may be that these features of infant development have adaptive significance. A vervet or a De Brazza's infant that stumbles or in any other way is clumsy will come to no harm if it is on or near the ground. It can, therefore, move away from its mother without risk. If an arboreal infant stumbles, it is in danger of falling out of its tree, possibly from a considerable height. Even if it didn't kill itself by doing so, it might be impossible for the mother to find or retrieve the infant again. It is clearly an advantage for the infant to remain within easy reach of the mother so that she can retrieve it rapidly when it is in danger of falling.

The close similarity of the figures for vervets and De Brazza's to those for the predominantly terrestrial baboons and rhesus, to which they are only distantly related, rather than to the arboreal mangabeys to which they are more closely related, suggests that this adaptive interpretation may be the correct one.

It should be added that the high figure for "mother holds" recorded for one Sykes mother and for the De Brazza's mothers was a response to frequent attempts by other adult females in their groups to steal their infants, and does not, therefore, reflect differences in habitat.

CHANGES IN MOTHER-INFANT BEHAVIOR FOLLOWING CHANGES IN GROUP COMPOSITION

It has been reported, both for captive and wild monkeys, that the behavior of a mother toward her infant may be greatly influenced by the behavior of the other members of her group. For example, Hinde *et al.* (1964) have shown that the presence of a possessive "aunt" within a group of *Macaca mulatta,* who continually tried to steal babies from their mothers, considerably increased the restrictiveness of the mothers toward their infants. Another case relevant to this article concerned a captive mangabey (*Cercocebus albigena*) mother living at the time with its own three-month-old baby, two adult females, and an adult male (Chalmers and Rowell, 1970). The adult male died of pneumonia and the mother, after his death, became extremely restrictive toward her own infant, holding it firmly for the entire day, and preventing the baby from leaving her, even though the latter struggled violently to do so. This suggested that the male, when alive, influenced the mother's behavior toward her baby, possibly reassuring her by his presence, or possibly in a more indirect way through his influence on the rest of the group.

It seemed important to discover whether this effect of the adult male's presence was peculiar to this one mother, or whether it occurred in other mothers of other species. It was particularly interesting to see if species that differed in their social organization and behavior would exhibit this effect to the same degree. An obvious comparison to be made here was between the patas monkey, *Erythrocebus patas,* and a member of the closely related *Cercopithecus* genus. Hall (1965) has shown that wild patas live in groups containing only one adult male who plays a very peripheral and independent role from the rest of the group. By contrast, reports by Struhsaker (1967a), Bourlière *et al.* (1970), and others on a number of *Cercopithecus* species indicate that here the adult males seem to be more closely integrated with the rest of the group. The Sykes monkey (*C. albogularis*), being the most readily available *Cercopithecus* at the time of the experiments, was used for the comparison.

The experiments to be described here were also designed to see if the reassurance effect described above was specific to adult males.

ANIMALS AND CONDITIONS OF CAPTIVITY

Sykes and patas monkeys were kept in monospecific groups, each of which contained one adult male, between two and five adult females, and their young. They were observed in outdoor cages, the dimensions of which varied slightly, some being 8.5 × 3.5 × 4.5 meters, others being 11.3 × 4.5 × 2.5 meters. All adults had been kept in captivity for at least three years, the majority for more than five years. All the young had been born into captivity. The groups had remained relatively unchanged in their composition over this period, but minor changes in group composition had been made four months prior to the present experiment

TABLE 3. Details of Infants Used in Male Removal Experiment

	NAME	SEX	PARITY OF MOTHER	AGE OF INFANT (DAYS)
Sykes	Jinx	M	Unknown	441
	Derby	M	Unknown	321
	Unnecessary	F	Primiparous	245
	Azel	M	Multiparous	116
Patas	Lennon	M	Primiparous	324
	Twinkle	F	Multiparous	220
	Chocolate	M	Multiparous	139
	Mary MacGregor	F	Multiparous	97

to obtain the desired group compositions. Following this, the groups had remained undisturbed. Four Sykes mothers from three separate groups together with their infants, and four patas mothers from two different groups together with their infants were watched. Table 3 gives the name, sex, and age for each infant used in the experiment together with the parity of their respective mothers.

METHODS OF OBSERVATION

Half of the mother-infant pairs were observed according to the following schedule:

1. The group was watched undisturbed for several days, one hour per day to habituate the animals to the observer.
2. The adult male was removed from the group. For the next five days the behavior of the mother and infant was recorded, one hour per day at the same time each day.
3. At the end of the fifth day the male was returned to the group. The behavior of the mother and infant was again recorded, one hour per day for the next five days.
4. At the end of this period an adult female (not the mother) was removed from the group. The behavior of the mother and infant was recorded one hour per day for the next five days.
5. At the end of this period the adult female was returned to the group, and five more hour-long, daily watches carried out.

For the remaining mother-infant pairs, the female removal and replacement was carried out before the male removal and replacement as a control against habituation to the experimental procedure and behavioral changes due to the aging of the infant.

The behavior of the mother and infant while the male was absent could thus be compared with that after the male had been returned, and similarly for the adult female. The reaction toward the absence of the male could also be compared with that toward the absence of the female.

Behavior was recorded on check sheets similar to those described earlier in this article. Two-thirds of the observations were carried out by the author, the remainder by trained assistants. Differences between measures for the different experimental situations were tested by x^2 with a 5 percent level taken as significant.

RESULTS

Time Spent by the Infant Away from the Mother. Table 4 shows the number of half-minute periods in each experimental situation that the Sykes and patas infants spent the entire half-minute period more than two feet away from the mother. With the Sykes, all except the oldest infant show a significantly lower number of half-minute periods totally off the mother with the male absent compared with the male being present. Only one infant shows a similar trend when the female is removed from the group; for the two other infants the figure for the time when the male is absent is significantly lower than the figure for the time when the female is absent.

With the patas no consistent trends are detectable. Moreover, in only one of the four infants (MM) is the figure for the time when the male was removed significantly different from that when the female was removed, and in this case the difference between the corresponding control periods is very wide. This suggests that the differences in results for the four different situations have little to do with the experimental manipulation. Why such large fluctuations should occur between the control levels is not known.

Maintenance of Proximity Between Mother and Infant. While the reduction in time spent away from the mother by the Sykes infants in the absence of the male could be due to an increase in the tendency of both mother and infant to remain close to each other, it is possible that one of the pair, either the mother or the infant, could be primarily responsible for this lower figure. If it is due to an increase in the tendency of the infant to stay close to the mother, then we would expect, on commonsense grounds, the proportion of approaches between the mother and infant that were due to the infant to increase and the proportion of leavings between the two due to the infant to decrease. In either case, the measure of the percentage of approaches by infant minus the percentage of leaves by infant (hereafter abbreviated $\%AP_i - \%L_i$) would increase. Conversely,

TABLE 4. Number of Half-Minute Periods in Which the Infant Spent the Entire Half Minute More Than Two Feet Away from the Mother (Five Hours Observation for Each Situation)

	INFANT	MALE REMOVED	MALE REPLACED	CHANGE	FEMALE REMOVED	FEMALE REPLACED	CHANGE
Sykes	Jx	506	454	*	436	435	N.S.
	D	403	465	†	450	440	N.S.
	U	249	299	†	202	336	†
	A	199	399	†	377	339	N.S.
Patas	Ch	247	369	†	273	310	N.S.
	MM	183	195	N.S.	275	300	N.S.
	L	360	416	†	384	502	†
	T	320	292	*	324	287	*

* : "removed" figure significantly higher than "replaced" figure.
† : "replaced" figure significantly higher than "removed" figure.
N.S.: "removed" and "replaced" figures not significantly different.
NOTE: These symbols indicate directions of change and are not indications of statistical significance.

TABLE 5. Percentage of Approaches Between Mother and Infant Due to Infant Minus Percentage of Leaves Between Mother and Infant Due to Infant (Five Hours Observation for Each Situation)

	INFANT	MALE REMOVED	MALE REPLACED	CHANGE
Sykes	Jx	17.3	24.8	†
	D	12.8	25.1	†
	U	42.2	18.7	*
	A	−1.1	8.3	†

* : "removed'" figure significantly higher than "replaced" figure.
† : "replaced" figure significantly higher than "removed" figure.
NOTE: These symbols indicate directions of change and are not indications of statistical significance.

if the reduced time spent by the infant at a distance from the mother were primarily due to an increase in the tendency of the mother to stay close to the infant, we would expect a decrease in the measure $\%AP_i - \%L_i$. For a full discussion of this see Hinde (1970). (Approach and leave here are defined as crossing, in the appropriate direction, a circle two feet in radius centered on the mother.)

Table 5 shows the value of $\%AP_i - \%L_i$ for Sykes when the male is removed and when it is replaced. It shows that in two out of three infants for which removal of the male resulted in a reduction in the time spent by the infant totally off its mother, it also resulted in a decrease in $\%AP_i - \%L_i$. This indicates, then, a change in the mother's behavior, an increase

TABLE 6. Number of Half-Minute Periods in Which Mother Held Infant (Five Hours Observation for Each Situation)

	INFANT	MALE REMOVED	MALE REPLACED	CHANGE	FEMALE REMOVED	FEMALE REPLACED	CHANGE
Sykes	Jx	1	5	N.S.	3	1	N.S.
	D	0	2	N.S.	0	0	N.S.
	U	52	20	*	9	6	N.S.
	A	199	29	*	9	18	N.S.
Patas	Ch	0	0	N.S.	7	4	N.S.
	MM	61	13	*	35	33	N.S.
	L	0	0	N.S.	0	0	N.S.
	T	0	0	N.S.	0	0	N.S.

N.S.: "removed" and "replaced" figures not significantly different.
* : "removed" figure significantly higher than "replaced" figure.
NOTE: These symbols indicate directions of change and are not indications of statistical significance.

in her tendency to stay close to the infant when the male is removed. With the third infant it was the infant who played a greater role in maintaining proximity when the male was removed. (Hinde and Atkinson [1970] have shown that the value of $\%AP_i - \%L_i$ can fluctuate with the level of activity of the infant, while the degree to which the mother and infant are responsible for maintaining proximity to one another remains constant. In no case in this experiment did the infant's level of activity fluctuate in a way to invalidate these results.)

Mother Holding the Infant. A measure of the restrictiveness of the mother is given by the number of half-minute periods in which the infant was held by the mother. Table 6 shows this measure for Sykes and patas. In the Sykes the mothers of the two younger infants held their infants significantly more often when the male was removed from the group than when it was replaced, and also more than when the female was removed. The mothers of the two older infants did not do so. No significant changes in holding occurred when the female was removed. In the patas one mother out of four held her infant significantly more when the male was removed compared with when the female was removed, and also compared with when the male was replaced.

DISCUSSION

These results suggest that removal of an adult male Sykes from the group results in an increased restrictiveness of the mother toward her baby, provided it is not more than about a year old, whereas removal of an adult male patas from the group does not have the same effect. Two possibilities suggest themselves as to why removal of the male should affect the mother's behavior toward her infant. One is that removal of the male disturbs the mother, so that she reacts in a more cautious and protective manner toward her infant than she would do when relaxed. Another possibility is that removal of the male upsets the social equilibrium of the group, dissolving dominance-dependent relationships (Imanishi, 1960) under which the mother might have gained protection from the adult male's presence.

To determine whether the latter was the case, records were kept during the infant watches of all interactions between any member of the group and the mother and her infant. If the mother's increased restrictiveness during the absence of the male were due to a decrease in her status, we would expect changes in behavior of other members of the group toward her during the absence of the male. Changes that might be expected are an increase in aggression toward the mother, an increase in attempts to steal the infant, an increase in the number of times the mother avoids other members of the group, and a change in the amount of time the mother grooms and is groomed. (Some reports indicate that the frequency of grooming is directly correlated with dominance status, others that it is universely correlated. For a review of the reports, see Sparks [1967]. Most authors agree, however, that there is a change in one direction or another, and we would therefore expect the measure—the time mother grooms individual A plus total time mother is involved in grooming bouts with A—to fluctuate if the mother's status varies at different stages of the experiment.)

The results obtained show that the amount of aggression toward the mother, the amount of avoiding by the mother, and the amount of infant stealing were at all times negligible, these behavior patterns rarely occurring more than once every five hours. Figure 4 shows grooming bouts between mothers and other females in their groups. It shows the percentage of times that the mother was involved in grooming with particular individuals in which she was the groomer. Cases in which the mother was subordinate to the other female are asterisked (status being assessed by grooming, feeding, and supplanting relationships). There are no consistent changes in grooming figures that could possibly indicate changes in status of the mother with removal of the adult male.

In the absence of evidence that the mother's status changes with removal of the adult male in a way that might cause her to become more restrictive toward her infant, it seems reasonable to assume that the mother's restrictiveness is a direct response to the removal of the male, or, put another way, that the presence of the adult male allows the mother to behave in a calm and relaxed manner.

Figure 4. Percentage of Time Sykes Mothers Engaged in Grooming Bouts in Which They Were the Groomers

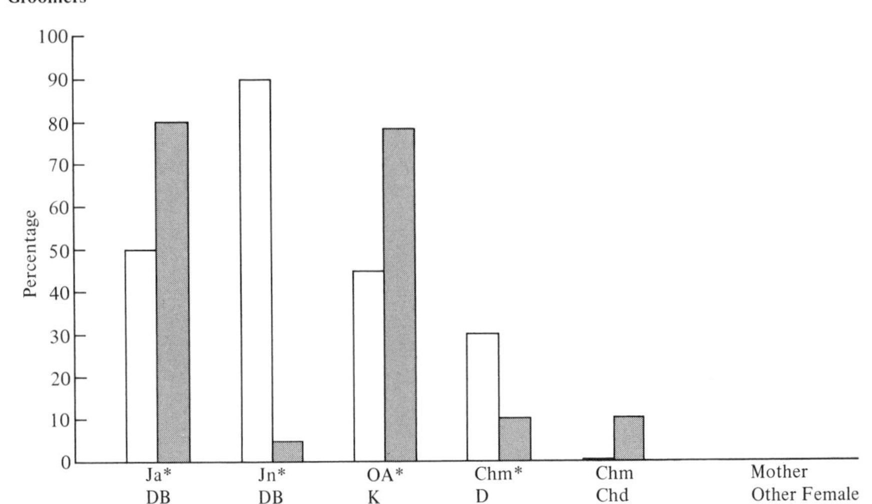

Unshaded columns, male removed. Shaded columns, male replaced. Columns represent the four mothers with five different females. Grooming between mothers and the remaining females was too infrequent to be meaningful. Asterisks indicate that the mother was subordinate to the other female..

Why this effect should be clear in Sykes but not in patas is open to speculation. Two possibilities come to mind. One is that the patas mother-infant relationship develops much more rapidly than that of the Sykes. If this were the case, patas mothers and infants would be less likely to be disturbed by the manipulation in group composition carried out than Sykes mothers with infants of comparable age. Casual observation on the rate of development of the two species does not support this. Moreover, the data presented in this article show that the youngest patas infant does not approximate more closely to the behavior of the Sykes infants than do the older patas infants.

A second possibility is that these differences in results reflect genuine differences in social behavior of the two species. As stated earlier, patas live in one-male groups in which the adult male plays a very independent role from the other members of the group, acting mostly as a lookout for the group, but interacting rarely with the adult females. Hall *et al.* (1965) have also shown that the females in captivity are by no means subordinate to the male, and may harass and attack him. That the absence of such an adult male from a captive group might not upset the mothers is not surprising.

Comparable data on the social organization and behavior of Sykes monkeys are not available. Reports by Aldrich-Blake (1970) on the very closely related blue monkey (*C. mitis*) suggest a close similarity to other arboreal *Cercopithecus* species; that is, it has loosely knit groups with an average size of fourteen, in which there is usually only one mature male, the male playing a leading role in the group's behavior. The absence of such a male from a captive group of Sykes may well be expected, therefore, to have an effect on the behavior of mothers toward their infants.

From what has been described in this and the first half of the article, it is clear that the nature of the habitat in which a species is commonly found, and also the structure and behavior of the group (themselves adapted to the habitat) influences the behavior of mothers and infants within the group. Clearly, the effects described here merely point to the sorts of explanations that lie behind the differences in mother-infant behavior in different species. One imagines that there must be a host of finer differences between species than the ones described here, explicable in terms of differences in habitat and social organization of the species.

SUMMARY

This article attempts to relate differences between species in mother-infant behavior to differences in their ecology, social structure, and social behavior. Certain aspects of the development of infants of four captive cercopithecine species, *Cercopithecus aethiops, C. neglectus, C. albogularis* and *Cercocebus albigena,* over the first three months of life are described. Infants of the more arboreal of these species tended to spend less time away from their mothers than did infants of comparable age of the more terrestrial species. This was due to a greater reluctance of the arboreal infants to leave their mothers, for they matured at the same rate as the infants of the more terrestrial species, and their mothers were no more restrictive than those of terrestrial species.

An experiment is described showing that the presence of the adult male in Sykes groups, but not in patas groups, appears to have a calming influence on the mother, reducing the level of restrictiveness toward their infants. The difference between the two species may be explicable in terms of differences in social organization.

Acknowledgments

This work was supported by funds from the L. S. B. Leakey Foundation, and from the Makerere University College Research Fund.

References

ALDRICH-BLAKE, E. P. G. "Problems of Social Structure in Forest Monkeys." In J. H. Crook (ed.), *Social Behaviour in Birds and Mammals.* London: Academic Press, 1970, 136–158.

BOURLIÈRE, F., C. HUNKELER, and M. BERTRAND. "Ecology and Behavior of Lowe's Guenon (*Cercopithecus campbelli Lowei*)." In J. R. Napier and P. H. Napier (eds.), *Old World Monkeys: Evolution, Systematics and Behavior.* New York: Academic Press, 1970, 297–351.

CHALMERS, N. R. "Group Composition, Ecology and Daily Activities of Free-living Mangabeys in Uganda." *Folia Primat.* 8 (1968), 247–262.

——— and T. E. ROWELL. "Behavior and Female Reproductive Cycles in a Captive Group of Mangabeys." *Folia Primat.* (1970).

DEVORE, I. "Mother-Infant Relations in Free-ranging Baboons." In H. L. Rheingold (ed.), *Maternal Behavior in Mammals.* New York: Wiley, 1963, 305–335.

FITZGERALD, A. "Rearing of Marmosets in Captivity." *J. Mammal.* 16 (1935), 181–188.

GARTLAN, J. S. and C. K. BRAIN. "Ecology and Social Variability in *Cercopithecus aethiops* and *C. Mitis.*" In P. Jay (ed.), *Primates: Studies in Adaptation and Variability.* New York: Holt, Rinehart and Winston, 1968, 253–292.

HALL, K. R. L. "Behavior and Ecology of the Wild Patas Monkey, *Erythrocebus patas,* in Uganda." *J. Zool. Lond.* 148 (1965), 15–87.

———, R. C. BOELKINS, and M. J. GOSWELL. "Behavior of *Erythrocebus patas* in Captivity, with Notes on the Natural Habitat." *Folia Primat.* 3 (1965), 22–49.

HINDE, R. A. "Analyzing the Roles of the Partners in a Behavioral Interaction —Mother-infant Relations in Rhesus Macaques." *Ann. N.Y. Acad. Sci.* (1970), in press.

——— and S. ATKINSON. "Assessing the Roles of Social Partners in Maintaining Mutual Proximity, as Exemplified by Mother-Infant Relations in Rhesus Monkeys." *Anim. Behav.* 18 (1970), 169–176.

———, T. E. ROWELL, and SPENCER-BOOTH. "Behavior of Socially Living Rhesus Monkeys in Their First Six Months." *Proc. Zool. Soc. Lond.* 143 (1964), 609–649.

IMANISHI, K. "Social Organization of Subhuman Primates in Their Natural Habitat." *Curr. Anthrop.* 1 (1960), 393–407.

JAY, P. "Mother-Infant Relations in Langurs." In H. L. Rheingold (ed.), *Maternal Behavior in Mammals.* New York: Wiley, 1963, 282–304.

MOYNIHAN, M. "Some Behavior Patterns of Platyrrhine Monkeys. I. The Night Monkey (*Aotus trivirgatus*)." *Smiths. Misc. Coll.* 146 (1964), 1–84.

ROWELL, T. E., N. A. DIN, and A. OMAR. "The Social Development of Baboons in Their First Three Months." *J. Zool. Lond.* 155 (1968), 461–483.

SAUER, F. E. G. and E. M. SAUER. "The South-West African Bush-baby of the *Galago senegalensis* Group." *Jl. S. W. Africa Scient. Soc.* 16 (1963), 5–35.

SIEGEL, S. *Nonparametric Statistics for the Behavioral Sciences.* New York: McGraw-Hill, 1956, 271.

SPARKS, J. "Allogrooming in Primates: a Review" in D. Morris (ed.), *Pri-*

mate Ethology. London: Weidenfeld & Nicholson, 1967, 148–175.
STRUHSAKER, T. T. "Behavior of Vervet Monkeys (*Cercopithecus aethiops*)." *Univ. of Calif. Publ. Zool.* 82 (1967a), 1–74.
———. "Ecology of Vervet Monkeys (*Cercopithecus aethiops*) in the Masai-Amboseli Game Reserve, Kenya." *Ecology* 48 (1967b), 891–904.
VAN LAWICK-GOODALL, J. "The Behavior of Free-living Chimpanzees in the Gombe Stream Reserve." *Anim. Behav. Monogr.* 1 (1968), 161–311.
WOOLDRIDGE, F. L. "Behavior of the Abyssinian Colobus Monkey, *Colobus guereza,* in Captivity." University of South Florida: Unpublished M.A. thesis, 1969, 78–82.

Jane B. Lancaster

Play-mothering: The Relations Between Juvenile Females and Young Infants Among Free-ranging Vervet Monkeys*

INTRODUCTION

Until the studies of the past decade on primate social behavior became available, it seemed to many that the major behavior patterns of a species, such as mating or maternal behavior, were largely instinctive. It was assumed by most scientists that vital activities like reproduction could not be left to the chances of a learning process and that such patterns were probably relatively fixed genetically. Recent field and laboratory workers have shown that in many species of mammals, and especially in monkeys and apes, learning and experience play vital roles in the development of the behavior patterns used in mating and maternal care. For example, Harlow and Harlow (1965) have shown that monkeys who have been raised in cages with only their mothers and deprived of chances to play with their peers are very ineffective in mating. Their motor patterns are poorly coordinated and directed, and their motivation is very low. As adults these monkeys can be given a kind of encounter group therapy that helps improve their mating performance, but it is still unlikely that they will develop into really effective adult animals. There is also an indication that there is an optimal period for learning behavior patterns such as copulation. This optimal period occurs before puberty. During the preadolescent period young animals spend much of their time in play groups in which they perform many of the patterns of mating and aggression that they will use in their adult lives.

These findings have encouraged a new focus for research on the juvenile period of development as a critical time for the establishment of patterns to be used in adulthood. The play of juveniles not only gives them opportunities to practice their motor skills but also contributes to the establishment of emotional attitudes that are essential to fulfilling their adult

* An earlier version of this article appeared in *Folia Primatologica* 15 (1971), 161–183. Reprinted by permission.

roles. This new point of view helps us to look at learning and experience from a biological perspective. It is clear that for many species besides man experience is important in the development of the major classes of behavior that are needed for survival and reproduction. This learning must occur for the individual to be a normal adult, and much of this learning and experience occurs in the context of juvenile play. Since it is clear that juvenile monkeys gain experience in mating, self-defense, and aggression through play, it is logical to wonder whether juvenile females gain experience in maternal behavior patterns before they become mothers themselves. Monkey and ape females have relatively few offspring during their lifetimes. Most do not mate until their third year of life or even later, and the long gestation combined with annual breeding patterns and single births make the loss of an infant through neglect or inexperience very costly.

RELATIONS BETWEEN INFANTS AND OTHER GROUP MEMBERS IN MONKEYS AND APES

A number of field workers have reported that in many species of monkey and ape the small infant acts as a social magnet for other group members. The mother with her new infant forms the center of a cluster of interested group members, and she and her infant are especially attractive to the other females of the group. In some species, such as the baboon (DeVore, 1963) and the Japanese macaque (Sugiyama, 1965), this orientation toward a new infant may be limited to peering at or trying to touch the infant. In other species the mother may permit group members to hold and carry her infant, as in the langur (Jay, 1962; Poirier, 1968; Sugiyama, 1965), the vervet monkey (Gartlan, 1969; Struhsaker, 1967b), and the chimpanzee (van Lawick-Goodall, 1968). In a recent summary, Hamburg (1969) has emphasized that the handling of infants by juvenile and adolescent females is common in monkeys and apes; a preadult female that is truly naïve with respect to maternal behavior is probably very rare in the wild.

A similar interest in infants has been reported in laboratory studies of a rhesus monkey colony (Hinde and Spencer-Booth, 1967; Rowell *et al.,* 1964; Spencer-Booth, 1968). Hinde and his associates found that in all their social groups some females played maternal roles toward infants of other females. This greatly affects the social experience of the young infant as it matures.

Nonhuman primate males may also display protective and affectionate responses toward young infants, but this behavior seems to be much more variable in both pattern and extent (DeVore, 1963; Chalmers, 1968;

Lahiri and Southwick, 1966; Mitchell, 1969). For example, Itani (1959) reports a type of annual adoption of yearlings by dominant males at the time when the females are giving birth to new infants. This behavior occurs in high frequencies in some groups of Japanese macaques, whereas in others it is rare or absent. Mitchell (1969) has recently summarized the evidence for parental behavior displayed by nonhuman primate males. This care is rarely directed to the very small infant before the coat changes to the adult coat color. When a nurturing or protective role is played by a male, he is likely to be a subadult or a fully adult male and sometimes a fairly high-ranking one. Such roles have not been reported for young, low-dominant males except in the special situation where a young male is protective of his younger sibling (Kaufmann, 1967; Koford, 1963; van Lawick-Goodall, 1968; Sade, 1965, 1967).

This article is particularly concerned with maternal, or protective and nurturing, behavior displayed by juvenile females toward infants. This behavior is of particular interest because it first occurs long before puberty has been reached and thus must reflect some very early sex differences in orientation toward the playing of a maternal role. Only a few field studies specifically discuss the attraction of infants for juvenile females as opposed to subadult and adult females, but this behavior is mentioned briefly by DeVore (1963), Anthoney (1968), and Bolwig (1959) for baboons; by Jay (1962) for langurs; by Struhsaker (1967b) for vervets; and by Altmann (1965) and Kaufmann (1966) for rhesus macaques. Such behavior can be inhibited very easily by tension in the group or even by the mild anxiety raised by the presence of an observer. At Cayo Santiago, for example, Altmann (1965) noted that juvenile female rhesus showed strong interest in infants, but since adult females were so highly protective of their infants, the juvenile females were rarely permitted to hold or even touch them. Altmann's observations were originally made when the colony was rapidly expanding in numbers and tension was high. But only a few years later at Cayo Santiago, Kaufmann (1966) often saw juvenile females holding and carrying infant siblings in the birth season. Wilson (p.c.[1]) reports that this behavior is now common there but that it is usually seen in the evening after most of the workers and technicians have left the island for the night. It is reasonable to expect that a mildly anxious mother monkey would tend to be more restrictive of her infant than a relaxed mother would be. She would be less likely to hand it over to a juvenile female who is inexperienced and clumsy with infants. Even though other kinds of behavior such as play or copulation may not be inhibited by very low levels of anxiety or tension, it is possible that the mothers' permissive-

[1]p.c.: personal communication.

ness toward the handling of infants by juveniles would be. This may be a kind of behavior that is often suppressed by the mere presence of an unfamiliar observer. As more field studies are done on animals that are highly habituated to an observer, caring for infants by juvenile females may emerge as a very common behavior pattern in nonhuman primates.

BEHAVIOR OF FREE–RANGING VERVET MONKEYS

ECOLOGICAL SETTING AND SOCIAL BEHAVIOR

The main study area was along the Zambezi River near Victoria Falls, Zambia. The altitude is about 3,000 feet with annual rainfall of twenty-four to twenty-eight inches. The climate is hot and dry with the rainfall concentrated in four to six months of the year. The rainy season begins in November and extends through April with the heaviest rainfall in January and February. The home ranges of the vervet monkeys included two major types of vegetation. The first was gallery forest that formed a belt rarely more than one hundred yards wide along the banks of the Zambezi River. This belt was composed of trees typical of wet conditions such as the African ebony, fig trees, the African mangosteen, and other fruit trees. The second type of vegetation was MuPani woodland, a savanna type of vegetation of deciduous trees with grass or open woodland that surrounded the gallery forest along the Zambezi. In the woodland area the streams were intermittent.

The main study group of vervet monkeys (*Cercopithecus aethiops*) controlled a home range that was composed of a two-mile strip along the river extending into the MuPani woodland for about a mile. The vervets used both types of vegetation extensively but the use varied seasonally. The monkeys tended to sleep along the river in the high trees, but sometimes they would stay in the woodland overnight. The vervet groups in the area were usually antagonistic when they met. The main study group always expelled neighboring groups that intruded into its territory. Nevertheless, because of the large size of the territories, intrusions were frequent, and there were large overlapping areas in the home ranges of neighboring groups (sometimes as much as one-half mile).

A total of 840 hours of observations of the social behavior of the main study group were made between July 1, 1968, and March 1, 1969. During the 1968–1969 birth season fourteen infants were born to the group of forty-one monkeys. The composition of the group, including the births of these new infants, is shown in Table 1.

The adult monkeys were organized into male and female dominance hierarchies. Although some females were dominant over some adult males,

TABLE 1. Age and Sex Composition of the Old Drift Group of Fifty-five Vervet Monkeys near Livingstone, Zambia

	MALES	FEMALES
Adult	7	15
Subadult (three year olds) *	1	3
Juvenile two (two year olds)	4	3
Juvenile one (one year olds)	4	4
Infant	9	5
Total	25	30

* Includes three females still nulliparous at the end of the birth season.

in many social situations the two hierarchies seemed to function separately. For example, in feeding experiments, subordinates waited their turn while dominants fed, but representatives of the male and female hierarchies fed side by side without regard to their relative ranks. Juveniles and infants ranked in dominance just below their mothers. The outcomes of many dominance encounters were heavily influenced by coalitions, potential or actual, based on mother-offspring and sibling relationships. Fully adult males were more active than any other age-sex class in territorial defense, in alertness for neighboring groups or predators, and in threatening predators. Copulations were relatively infrequent (only twenty-two observations over an eight-month period) and mostly confined to the mating season. Most social interactions were peaceful, and fights and chases were rare although at least one episode a day was not uncommon.

PLAY-MOTHERING

Among vervet monkeys living near the Zambezi River in Central Africa, juvenile females are often seen carrying, holding, or grooming infants. This behavior is seasonal because vervets in this area give birth to infants only during a four-month period each year. The first infants are born in late October, just before the beginning of the rainy season, and more are born in November, December, and January. With vervets it is the small black infant that gets the most attention. During the first three months of life the black infant's social contacts are restricted. It spends its time in close proximity with its mother, its older siblings, and juvenile and adolescent females from other genealogies. By the time the infant is three months old, its coat changes from the natal black to the light gray adult color. At about the same time the mother begins to refuse to carry the infant unless there is some immediate difficulty or danger. At this same age infants also begin to enter the juvenile play group and to form social bonds with other young animals and with adult males.

The onset of the birth season marks an abrupt change in the behavior of juvenile females. Before the infants are born each year, juvenile females spend most of their time either in the company of their mothers and siblings or in play groups with other juveniles. However, during the months when there are young black infants in the group, the juvenile females have a new focus for their interest that draws them away from their family groups and from play with each other and with young males. The small black infant acts as a magnet for a juvenile female regardless of her age or social position. It is common to see each new mother acquire an entourage of juvenile females, who follow her about during the day waiting for a chance to touch the infant. Even infant females who are only nine months old and not yet weaned themselves show great interest in newborns. Young females will spend as much time with an infant as they possibly can, and it is the mother's attitude that usually determines how long the juvenile females will be able to handle an infant.

This great attraction of the young infant for juvenile and adolescent females has been observed in vervet populations in two other localities. Gartlan (1969) working on Lolui Island in Lake Victoria observed maternal responses to infants displayed by juvenile and subadult females during the early period before the infants' color change. He also emphasized that adult and subadult males were never observed to show interest in small infants (Gartlan and Brain, 1968). Struhsaker (1967b), who worked in Amboseli Reserve, Kenya, reports similar behavior in his monograph, and his films contain a number of sequences showing subadult or large juvenile females carrying or holding infants.

Vervet mothers are normally relaxed about their infants. Vervets do not go to the extreme of passing infants around among all the adult females of the group as do langurs (Jay, 1963; Poirier, 1968; Sugiyama, 1965). Nevertheless, a vervet mother will often let another female come and sit nearby and touch or hold the infant as long as she remains near the mother. Juvenile female vervets show far greater interest in infants than do the adult females. A total of 347 observations of affectionate contacts between females and infants not their own was made between October 12, 1968, and February 28, 1969, during a total of 464 observation hours. Only 52 of these contacts were made by adult females, that is, females who were three years old or older and who had had infants either in previous years or during the birth season of 1968–1969. There were 295 contacts between infants and nulliparous females aged one, two, and three years old. Excluding infants, the nine nulliparous females, only two of which were adolescent, composed 38 percent of the females, yet they accounted for 85 percent of the contacts with infants. (Data from a nulliparous three-year-old female have been discarded because she had a severely broken leg at the beginning of the birth season and was barely able to keep up with the group.) Furthermore, contacts with infants by juvenile females tend to be more sustained than do those of adult females. Typically an adult female will stride over to a mother with a new infant and greet the infant by briefly nuzzling or sniffing its head. She may then reach down and pull the infant up by its tail or hindquarters and nuzzle its genitals. Because the infant maintains contact with its mother by clinging with its hands and mouth, it is raised up hingelike from its mother's body. After nuzzling the genitals, the adult female will abruptly let the infant go and then will direct her attention to grooming the mother, showing no more interest in the infant itself. Virtually identical behavior has been reported by Gartlan (1969) and by Struhsaker (1967b) for vervet monkeys living in East Africa.

During the first few weeks after the infant is born, the mother tends to restrict the infant's movements and to keep it close to her body. Juvenile females will follow the mothers of new infants and sit beside them and peer at or try to touch the infant. Often the juvenile female will begin by grooming the mother, only gradually, but very obviously to the observer, working her way over toward the infant. If the mother moves or seems disturbed by this attention to the infant, the juvenile female will hurriedly begin to groom the mother again. Similar behavior to this has been described by Hinde, Rowell, and Spencer-Booth (1964) for their rhesus colony, by Kaufmann (1966) for rhesus on Cayo Santiago, and by Gartlan (1969) for vervet monkeys in Uganda. This pattern contrasts with the behavior of adult females, who seem to groom the mother for the sake of grooming her and not to be near the infant, which is only briefly greeted by nuzzling or sniffing. During the early weeks of the infant's life the juvenile female will try to pull it away from its mother, or she will quickly scoop it up if the mother sets it down for a few moments. Usually the mother will take the infant back again fairly quickly, but the juvenile female will have a chance to hold the infant and briefly groom it or hug it to her chest. The youngest infant observed being held by a juvenile female was eight days old. By the time the infant is three weeks old, the mother is usually much more relaxed in her attitude toward it, and juvenile females will often be permitted to hug and carry it for quite some time.

When the very young juvenile females get their first chance to hold an infant or carry it, they often have difficulty in orienting its body properly or in instilling enough confidence in the infant to make it cling when the female tries to walk with it. Infants will cling readily to experienced females who are not their mothers, but they will often struggle when a young

TABLE 2. Mode of Return of Infant to Its Mother After Being Held by a Juvenile or Adolescent Female

Mother takes infant back	23
Mother presents for grooming	19
Mother grooms female	13
Mother threatens female	12
Infant returns by itself	19
Total	86

juvenile has them. Because the infant refuses to cling properly, it is common to see a juvenile female carrying an infant by walking along on three legs while clutching the infant to her chest with one hand. Sometimes a juvenile female will even hold an infant with both arms and run bipedally to try to get it out of its mother's view. Juvenile females soon learn how to carry infants properly, and they also learn that if they can keep the infant quiet and content the mother will probably not try to retrieve it. One of the best ways to pacify any monkey is by grooming it, and this works just as well with infants as with adults. A juvenile female is often seen pinning down an infant with her leg or arm and then intensely grooming it until it relaxes. In fact, this is such a common pattern that almost all an infant's grooming experience is with juvenile females rather than with its own mother, who usually only grooms it when it is obviously dirty or has something sticking to its skin or fur. The infant does not receive strictly social grooming from its mother but rather from the juvenile females.

A mother rarely has any difficulty in retrieving her infant when she wants it back. Table 2 summarizes the mode of return of the infant in the eighty-six such returns that were clearly observed. One common mode of retrieval is for the mother to meander up to the juvenile and present herself for grooming. The juvenile female will release her hold on the infant in order to groom the mother, and then the infant will usually return to its mother of its own accord. If the juvenile female fails to respond to the presentation or tries to ignore it by turning her back on the mother and hunching over the infant, the mother may simply circle around and present herself again until the infant is released. If this fails, the mother may begin to groom the juvenile female, which usually leads the juvenile female to release the infant as she relaxes and stretches out to enjoy the grooming. Often the retrieval does not involve any grooming at all. The mother simply walks over to the female, reaches out her hands to the infant, and pulls it away from the juvenile's arms. Occasionally the mother resorts to threats such as a mild head bob and stare. This may be necessary when an adolescent female gets hold of an infant and refuses to give it up. As the infant gets older, it can return to its mother itself when it tires of being with the juvenile females, and so retrieval becomes less of a problem.

The juvenile female usually plays a maternal or protective role toward the infant. Play is not observed between the infant and juvenile when the infant is very young. The earliest instance of such play was seen when an infant was thirty-six days old. As the weeks go by, social play in the form of chasing and wrestling gradually increases in frequency until it seems to replace the mothering behavior. By the time an infant is four months old and has finished its color change, it no longer seems to arouse maternal responses in juvenile females; instead it is seen as a social partner for play.[2] During July, August, September, and most of October, no maternal behavior by juvenile females directed toward infants was observed, although play sequences were common. By the time observations were ended on the study group in March, mothering behavior had already begun to decline in frequency, although there were still some small infants in the group.

By the time an infant is six or seven weeks old, it is actually spending a good proportion of its waking hours in the company of juvenile females. Mother vervets often take advantage of this and go off on their own to feed. Often when a mother vervet wants to feed out in the open or in a fruit tree with a number of other monkeys who may be excited and aggressive, she may leave her infant behind with some juvenile females. An infant may be left for as long as one-half hour or even an hour while its mother is feeding perhaps one-tenth of a mile away. Mothers will sometimes leave their infants with juvenile baby-sitters under other circumstances when bringing it along might be dangerous. For example, in one situation when a snake alarm call went up after the group had come across a python lying in the grass, a primiparous female stopped, pulled her infant from her chest, and set it down beside a juvenile female. She then ran over to the snake and joined the others in chattering at it while her infant remained about twenty feet behind with the juvenile.

For vervet infants this early period, when they are about two months old, is a time when potential loss to predators is high. As long as the

[2] It is not clear exactly how the changes in behavior and coat color are related. It is certain, however, that the development of the adult coat color marks a change in social status for the young infant. It is at this time that the infant first extends his social contacts beyond his mother, siblings, and the nursery group. He becomes the youngest member of the juvenile play group, and he also begins to interact with adult males for the first time. This usually involves nuzzling and touching an adult male's penis and scrotum and sitting next to or leaning against him during rest periods.

infant has company, it will not give a lost call but will wait patiently where its mother has left it. The juvenile baby-sitters are often very solicitous of infants left with them. They will follow them and watch their movements carefully. If an infant should have trouble moving about in a bush, a juvenile female will be quick to retrieve it or to pull it over near her. This solicitude toward infants is apparent in very young juvenile female vervets, even in those less than a year old. However, the juvenile baby-sitters sometimes lose interest in the infant and after one-half hour or so may move on with the rest of the group. The young infant will not follow them but will stay in the bush or bunch of trees where its mother left it. When the infant finds itself alone, it will give lost calls until it is retrieved. In two instances mothers were seen to answer lost calls by running back more than two-tenths of a mile to retrieve their infants, who had presumably suddenly found themselves alone. Lost calls are highly individualistic in form (Struhsaker, 1967a), and mothers appear to recognize the voices of their own infants. Mothers seem to be more restrictive of their infants when the group is moving through open country than through wooded areas. Gartlan (Gartlan and Brain, 1968) noted similar differences between his two study areas. At Chobi groups fed in dispersed patterns in open grassland, and the mother-infant bond seemed to be much stronger than at Lolui where home ranges were compact, population density high, and the vegetation provided protection against predators.

Table 3 summarizes the frequency of contacts between the nine nulliparous females in the group and the fourteen infants. The infants are listed according to their sex and age, for example, male 1 is the oldest infant. Female-infant pairs composed of animals known to belong to the same genealogy are marked by an asterisk. The differences between infants in frequency of contacts partly reflects the fact that some females, such as the mothers of Males 1, 2, and 5, tolerated the observer's presence much better than did others and hence were more willing to allow their infants to be held by other females when under observation. However, the amount of attention given to infants correlates most highly with how early in the birth season the infant was born. By the time the last infant (Female 5) was born, she attracted very little attention even though her mother was very habituated to the observer, and accordingly would not restrict contact with her infant out of anxiety. By that time, in late January, most of the juvenile females spent their time in large play groups with infants and male juveniles and were not following mothers with newborns. It may also appear from Table 3 that male infants get more attention than female infants, but it is doubtful that this is a true correlation. A larger proportion of females was born late in the season, and also three of the five mothers of females were unusually shy.

TABLE 3. Frequency of Affectionate Contacts Between Infants and Juvenile or Adolescent Females

INFANTS	ONE-YEAR-OLD FEMALES			TWO-YEAR-OLD FEMALES			NULLIPAROUS THREE-YEAR-OLD FEMALES			
	A	*B*	*C*	*D*	*E*	*F*	*G*	*H*	*I*	TOTAL
Male 1	21	8	8	1	4	3	3	6	3	57
Male 2	15	2	11	1	7	1	1	10	1	59
Female 1	4	3	3	1	—	—	4	3	7	25
Male 3	1	—	—	2 *	—	1	1	2	2	11
Female 2	6	5	4	1	3	—	—	1	1	21
Male 4	7	2	3 *	1	4	2	1	3	—	23
Male 5	11 *	8	8	2	1 *	3	2	7	6	48
Male 6	1	—	—	3	—	1	1	1	1	8
Male 7	1	1	3	—	—	—	1	1	3	10
Female 3	6	5 *	1	—	—	—	—	2	1	15
Male 8	3	1	—	—	—	— *	—	—	—	4
Female 4	—	—	—	1	—	—	— *	—	—	1
Male 9	—	—	—	—	—	—	—	—	—	—
Female 5	2	1	1	1	—	—	—	1	—	6
Unidentified	2	—	—	—	—	1	2	—	2	7
Total	80	36	44	14	19	12	16	37	27	295

* Individuals known to belong to the same genealogy.

No male of any age was seen to direct any maternal behavior such as hugging, carrying, or grooming toward a newborn infant. Fully adult males were tolerant of infants and the infants were attracted to them, but this occurred only after the infant was three months old or older. Juvenile males were never seen to show protective or maternal behavior to newborn infants, although older male siblings were seen to investigate their new siblings by touching or nuzzling them. However, only three infants were known to have juvenile male siblings, and two of these were born late in the field study, so that this type of behavior was rarely observed. Older male siblings oriented their behavior more often toward their mother than toward her infant during the early months. It was only when the infant became socially playful that juvenile males showed much interest in it.

There was no indication of any strong orientation of juvenile females toward their newborn siblings (Table 3). This is probably because all vervet females were relatively permissive so juvenile females would be attracted simply to any young infant. The yearling Female C, for example, was seen with four infants more frequently than with her own brother, Infant Male 4. In fact, she had only just been weaned herself, and her mother seemed to dislike the juvenile's attempts to pull the infant from the nipples. The mother several times threatened away her own juvenile while tolerating the approach of others.

Juvenile females varied greatly in the frequency with which they mothered infants. Female A was observed mothering infants more frequently than any other juvenile. She was the youngest of the yearlings and the daughter of a high-ranking female. Her mother was the only female who did not have an infant in the 1968–1969 birth season. The yearling was probably born at the very end of the previous year's birth season, and she was still seen sucking at her mother's nipples as late as February 1969. She directed most of her maternal behavior to three young males. One was Infant Male 5, who was most probably the son of her older sister. The other two, Infant Males 1 and 2, were the sons of females who ranked just above and just below her own mother in the dominance hierarchy. She was first seen to try to hug and carry infants when she was about nine months old and was hardly big enough to carry them. Her older sister, two-year-old Female E, showed much less interest in infants, a total of only nineteen contacts being observed. All three of the two-year-old females showed low frequencies of contact with infants in comparison to the yearlings and the adolescent three year olds. Whether this is an important difference linked in some way to ontogeny is not clear.

DISCUSSION

LABORATORY RESEARCH

Only one other study has been particularly concerned with maternal responses directed toward infants by nonmothers. This work has been done by Hinde, Rowell, and Spencer-Booth on a colony of rhesus monkeys (Hinde and Spencer-Booth, 1967; Rowell *et al.,* 1964; Spencer-Booth, 1968). They have reported on what they call "aunt" behavior, which includes affectionate, maternal, protective, and playful responses displayed by females of all ages toward infants. Much of the behavior they report is nearly identical to the behavior seen in free-ranging vervets. For example, just like vervets, female rhesus will seem to use subterfuge to get near an infant not their own. They will sidle up to the mother while appearing to forage for food, or they will groom the mother until her anxiety is lulled, and then surreptitiously groom the infant. These workers also observed young females showing anxiety and protective responses toward infants when they were in danger of falling from the walls of the cage or of being hit by a swinging door. Spencer-Booth (1968) found that nulliparous females, ages one through three, showed higher frequencies of this behavior (including touches, cuddles, and approaches) than did multiparous females, even on occasions when the experienced females did not have infants of their own. Spencer-Booth also noted that the maternal responses of nulliparous females were essentially the same regardless of their age, although two year olds showed slightly higher frequencies of such responses than did the other two age groups.

PREPUBERTY SEX DIFFERENCES IN BEHAVIOR

One of the most interesting questions raised by this data is whether there is a strong sex difference before puberty in the development of maternal behavior patterns. It has already been established in laboratory studies that sex differences in the behavior of rhesus monkey infants are readily apparent in such patterns as mounting and presenting, branch-shaking, and the roughness and duration of play bouts (Harlow and Harlow, 1965; Hinde and Spencer-Booth, 1967). Furthermore, clear differences have been found between male and female juvenile rhesus in the behavior they direct toward infants (Spencer-Booth, 1968). A special study on this problem was done by Chamove *et al.* (1967) in which they tested a series of fifteen preadolescent male-female pairs which had been raised under a variety of social conditions. Each monkey was put alone in a cage with a one-month-old infant. They found that the males showed ten times more hostility toward the infant than the females, whereas the females directed four times as much positive social behavior toward the infant. The juvenile females typically showed maternal, affiliative patterns such as hugging or grooming the infant, whereas the males were either indifferent or actively hostile (one male bit off an infant's finger). These sex differences were apparent in juveniles that had been raised with only a mother or with only a peer; however, they were absent in monkeys that had been reared in social isolation. These sex differences were the most marked in monkeys who experienced real monkey mothering.

Hamburg and Lunde (1966) have summarized some of the latest research in the development of sex differences in mammals. As an example of such research see the work of Goy (1969) and his associates on the rhesus monkey. Hamburg and Lunde note the high levels of hormones circulating in the blood of newborns. They suggest that during fetal or neonatal life hormones act in an inductive way on the undifferentiated brain to organize certain circuits into male and female patterns. Early exposure to these hormones may then affect the ease of learning and expression of the appropriate behavior patterns later in life, even though the level of sex hormones is very low during the period between infancy and adolescence. Hamburg and Lunde summarize a number of ways in which the hormones may act to produce the behavior patterns. For example, hormones at a critical period may affect later sensitivity to certain stimulus patterns. This may account for the differential behavior of young male and female vervets toward infants; it was mentioned earlier in this article that a number of field and laboratory studies have noted that males are usually indifferent to the infant in its black natal coat, whereas females are particularly attracted to an infant at that early stage of development. Hormones may act also by making the brain react in such a way that certain patterns of action are perceived as more rewarding than others. For example, the hugging of an infant to the chest may be very pleasurable to a female, whereas the large muscle movements and fast actions used in play-fighting and aggressive behavior may be felt as more pleasurable to a male.

Sex differences in behavior sometimes may be developed partly by the dynamics of social interaction within the group. As suggested by the laboratory studies of Chamove *et al.* (1967), animals with no social experience do not show marked sex differences in behavior. In another study, on pigtail macaques, Jensen *et al.* (1968) found clear sex differences in the development of independence from the mother. Almost identical sex differences were found in the development of infant rhesus monkeys by Mitchell (1968). These studies noted that both the mother and the infant showed differences in behavior that depended upon the sex of the infant. The mothers of males were more punishing and rejecting than the mothers of females, who tended to be restrictive and protective of their daughters. From the beginning of life male and female infants were treated differently by their mothers. The authors suspected that male infants behaved differently from females and that their mothers were in fact responding to this difference. Unfortunately the measures used in the study were too gross to pinpoint exactly what these primary differences were. It may be worth noting here that many field workers have observed that group members pay close attention to the genitals of newborn monkeys and apes. This behavior which includes peering at, touching, and sniffing or mouthing the genitals may represent a classifying of new group members as either male or female. As Benedict (1969) has suggested, even in animal societies social roles are not strictly inheritable, and it is through the interaction of individuals that social organization is produced. The way the individual learns roles, even male and female sex roles, may be just as heavily influenced by the environment of the moment as by the genetics of the individual.

PLAY-MOTHERING

The question remains whether juvenile female monkeys actually learn and benefit from experience in handling and caring for infants before they reach puberty. Although some of the basic patterns of maternal behavior may be relatively inborn, it is very likely that learning plays an important part in the development of skill in performing those patterns. It is clear from a number of field studies that young juvenile females are very inept in handling infants, but by the time a female reaches subadulthood, she carries and handles an infant with ease and expertise. Several field workers have reported their impression that they could see some development of skill in individual females as their experience with infants increased (Jay, 1962; Lancaster, this article; Struhsaker, 1967b). The dynamics of this learning process occur under the eyes of the real mothers. Instances of carelessness, clumsiness, or real abuse will, in effect, be punished. Normally, if anything should make an infant cry out, its mother will immediately come and retrieve it. If the infant is being abused, she may even bite the juvenile female. In this way through a simple kind of conditioning, juvenile females learn appropriate behavior patterns with their reward being the continued presence of the infant.

Two laboratory studies tend to support the idea that this early experience may in fact be practice for adult maternal behavior patterns. Seay (1966) found that he could not distinguish between primiparous and multiparous rhesus mothers that had been raised in the wild in respect to the mechanical aspects of skillful handling and caring for infants such as cradling, restraining, retrieving, embracing, and ventral and nipple contact. He found that primiparous mothers tended to be more anxious and sometimes more restrictive toward their new infants, but this in no way made them less effective mothers. In contrast experienced mothers were often more rejecting and punishing. In field studies done by Gartlan (1969) on vervets and by Kaufmann (1966) on rhesus no significant differences were noted between primiparous and multiparous females in their effectiveness as mothers. This does not mean that maternal behavior patterns are completely inborn. We know that a total lack of social experience does lead to the development of very infantile and aggressive mothers. Harlow *et al.* (1966) found that mothers raised in semi-isolation without mothers of peer relationships responded to their first infants with active rejection and hostility. However, they also found that social experience with an infant, however minimal and late in life, still affected maternal behavior. The same females who rejected their first infants often accepted their second ones. Although these females would probably not have made good mothers in the wild because of their lack of experience and skill in handling infants, their attitude toward infants clearly changed and they were able to take a nurturing and protective role toward their subsequent offspring.

It is clear that the opportunities to handle and care for infants before reaching puberty may play an important role in the development of maternal behavior patterns in the adult female. There are no data that suggest that primiparous females are less effective mothers than are multiparous ones, and it seems reasonable to assume that playing a mothering role as a juvenile may contribute to the success of the primiparous mother. Further laboratory studies will help to define some of the contributing factors and perhaps settle such problems as the effects of deprivation on the development of maternal behavior or the possibility of a critical period for learning maternal behavior patterns. However, the true test of adequate mothering can only come from studies on free-ranging animals. Adequate mothering in the laboratory may involve no more than passive acceptance of the offspring, whereas adequate mothering in a natural setting concerns a far wider range of problems including predation, safe locomotion over dangerous areas, and relations with other group members. It involves not only the skilled performance of maternal behavior patterns but also the motivation to play a maternal role and the ability to feel anxiety over the well-being of the infant.

ADAPTIVE VALUE TO THE INFANT

Gartlan (1969) has noted that maternal behavior by juvenile and subadult females may have adaptive value, since it contributes to the survival of the infant as well as to the development of maternal behavior patterns in the young females. He pointed out that within the genus *Cercopithecus* forest-dweling species do not have a contrasting coat color in infants. He suggested that in species where this contrast occurs, it may reflect a special vulnerability of small infants to the dangers of the environment. The contrasting color helps to mark out the small infant as an object of special interest and to keep it as the center of attention of at least one female in the group at all times. This would be especially important in species such as vervets in which the infants are large compared to the body size of the mother and are obviously a burden to the mother after they are six to eight weeks old. At present there is very little known about the behavior of species that have no contrasting natal coat, but it is possible that in those species the infant keeps in close body contact with its mother for a much longer period of its development.

Another adaptive value of prepuberty maternal behavior may arise in cases of adoption when an infant's mother dies. Obviously if the infant is too young to be weaned, it will not survive unless taken by a female capable of lactating. However, it is conceivable that, if the infant is two months old or more, it might survive under the protection of a subadult or juvenile. Observations of adoptions are rare in field studies but the few times they have been reported, juvenile animals have cared for the infant. In these adoptions, it has always been an older sibling that has attempted to take over the mother's role and not another adult female from the group. Sade (1965) reports the case of an infant rhesus who was protected and cared for by her older sisters, aged one and four years. The infant spent most of her time and grooming activity with her juvenile sister who was only one year older than she. Van Lawick-Goodall (1968) reports on three instances when a mother chimpanzee died and her infant was adopted by an older sibling. In two cases it was an older juvenile sister who cared for the infant, but in the third it was an older brother. The older sibling slept with the infant, groomed it, protected it, traveled with it, and sometimes carried it. In these observations of adoptions, few as they may be, it is interesting that it was not a fully adult female of the group who took the infant but one of the infant's close relatives, even though they were immature animals. The death of a mother is a much rarer occurrence than the death of an infant. However, the chances of survival of an orphaned infant would be much higher if it belonged to a line in which there were older siblings experienced in caring for infants. The adaptive value for the infant is probably secondary to the adaptive value for the juvenile who practices maternal patterns before she has her own offspring. However, both advantages work in the same direction and select for the early development of maternal behavior patterns.

MATERNAL BEHAVIOR AND PLAY

Play has always presented a problem when a comprehensive definition is needed. In a recent review Loizos (1967) has tried to bring together the published accounts of play in the higher primates. Most often play is described in terms of patterns relating to aggression and sexual behavior or, more rarely, to predator defense. Aside from man there seems to be no published account of play in mammals that involves maternal behavior patterns. This is somewhat surprising, since play is often mentioned as possibly serving the function of providing practice for behavior patterns important in adult life (Beach, 1945; Loizos, 1967; Welker, 1961). When we see a juvenile female of the human species display similar maternal behavior patterns toward a doll, we do not hesitate to call it play.

It is interesting that during the season when the juvenile and adolescent vervet females are preoccupied with small infants, they are conspicuously absent from the juvenile play groups. After the first bout of feeding every morning the group normally settles down for a rest. For adults this usually includes dozing or grooming either alone or with others, and for juvenile and subadult males this is a period of play-fighting and chasing. For juvenile females this is the best time to handle and groom infants. It is only after the infants have had their color change and have joined the play groups themselves that the juvenile females reappear with the male juveniles. Even then their primary orientation is first in romping with the infants and not with the juvenile males.

This behavior, in which juvenile females spend much more time caring for infants than play-fighting with their peers, may correlate with the dominance relations that seem to characterize the adult members of the group. Adult vervet females appear to be ranked in a linear hierarchy with offspring taking their dominance rank directly under their mothers.[3] From the field data it is clear that a juvenile female vervet can dominate any adult female that her mother can dominate even if her mother is absent at the time (Lancaster, this article). Dominance hierarchies described by Struhsaker (1967b) suggest that this is also true for vervets in Kenya. This same phenomenon has been reported for rhesus monkeys (Sade, 1965, 1967) and for Japanese macaques (Kawai, 1958; Kawamura, 1958). In these species on which long-term studies have been made and biological relationships are known, it is clear that a rank dependent on that of the mother carries over into adult life, at least for females. In contrast to females, male rhesus and Japanese macaques originally take their rank from their mothers, but at puberty a son can achieve rank above his mother through successful fights and threats, or by forming a coalition with another male. It is quite likely that the vervets follow a pattern similar to that of the macaques, although only long-term field studies can demonstrate whether this similarity is genuine or only superficial. If in fact female vervets do take their rank from their mothers even as adults, then it would be unimportant to a vervet female whether she is bigger, stronger, or a more skilled fighter compared with the other females in her group. Her dominance rank does not appear to relate to these qualities at all. Accordingly, play-fighting is far less important in her life, since

[3] In short-term field studies such as this one actual genetic relationships between individuals are hard to determine except for yearlings that have not yet been weaned at the time the study began. However, in a number of cases a particular two year old almost invariably slept, fed, and traveled with a particular adult female. From what is known about behavior of monkeys where genealogies are known such as at Cayo Santiago, it is reasonable to assume that the juvenile is closely related to the adult female and is probably her offspring or at least the offspring of her mother or sister. In addition, vervets that are closely related often show similarities in face and physique like those in human genealogies.

she does not need to practice highly the motor patterns and to develop skill in fighting, nor does she need to test herself against the other females. As adaptive preparation for adult behavior patterns it is appropriate that the young male vervet should spend his time play-fighting, while the young female spends much of hers playing with infants.

Loizos (1967) has suggested that play-fighting in primates may serve a function similar to that of imprinting in birds. It marks a period in which a young animal learns about the physical qualities of its conspecies and learns "which species it belongs to" (Loizos, 1967, p. 211). It may well be that juvenile female vervets also go through a period of "imprinting" on vervet infants while they are playing with them. The early appearance of maternal behavior patterns may serve two separate functions in the development of a juvenile female: first to help her to develop skill in caring for infants, and second to help her to learn to accept the role of mother toward an infant. Both of these functions are equally crucial to the survival of the infant, and it is not surprising that some practice in them should occur before the first infant is actually born.

This kind of data also raises questions about the development of maternal behavior patterns in humans. There is no doubt that if a human female is to be a truly effective mother, she must derive some kind of satisfaction from body contact with her baby. Nursing, cradling, and carrying must be positively motivated from within if the infant is to develop normally with a feeling of security. A mother who rejects her infant may go through the motions of caretaking, but the infant may respond to the way she does these patterns and react to the rejection. Perhaps more attention should be paid to what young girls are learning when they are playing with infant siblings or with their dolls, activities that seem to be universal. How important is this play in the development of maternal behavior patterns? Is it possible that there may be a kind of optimal period before puberty in which the emotional attitudes toward infants are first established? The ontogeny of maternal behavior patterns has been very poorly described or understood in our own species. We know something about cultural differences in ways that children are socialized in various societies but we know almost nothing about what important biological forces underlie the development of maternal behavior patterns in all human societies.

SUMMARY

Juvenile female vervet monkeys show a high degree of interest in young infants. They will touch, cuddle, carry, and groom infants whenever they have the opportunity. In a study of social behavior of vervet monkeys living along the Zambezi River near the Victoria Falls of Livingstone, Zambia, more that 295 observations were made in which a juvenile female directed some type of maternal behavior toward an infant. There is evidence from the literature that this kind of behavior may be common in many species of monkey and ape, but that the presence of an observer may greatly depress its frequency unless the group is highly habituated to man. This opportunity to care for infants provides juvenile females with situations in which they can practice not only motor skills that are important in maternal behavior but also playing the maternal role itself. This behavior raises interesting questions about the development and expression of sex differences before puberty, but further research is needed before the relative importance of experience and biological factors can be evaluated. It is possible that the care and handling of infants by juvenile female vervets may be looked upon as a variety of play behavior, one that is very similar to the human pattern of young girls playing with dolls or infant siblings.

Acknowledgments

Field work on which this paper is based was supported by a National Science Foundation grant (GS 1414) for the study of vervet monkey social behavior in Zambia, Africa.

I particularly want to thank Edna Brandt, Anne Brower, Phyllis Dolhinow, Chet Lancaster, Gary Mitchell, Thelma Rowell, and Sherwood L. Washburn for their many helpful comments on the first version of this paper.

References

ALTMANN, S. A. "Sociobiology of Rhesus Monkeys. 4. Testing Mason's Hypothesis of Sex Differences in Affective Behavior." *Behavior* 32 (1965), 49–68.

ANTHONEY, T. R. "The Ontogeny of Greeting, Grooming, and Sexual Motor Patterns in Captive Baboons (Superspecies *Papio Cynocephalus*)." *Behavior* 31 (1968), 358–372.

BEACH, A. F. "Current Concepts of Play in Animals." *American Naturalist* 79 (1945), 523–541.

BENEDICT, B. "Role Analysis in Animals and Men." *Man* 4 (1969), 203–214.

BOLWIG, N. "A Study of the Behavior of the Chacma Baboon." *Behavior* 14 (1959), 136–163.

CHALMERS, N. R. "The Social Behavior of Free-living Mangabeys in Uganda." *Folia Primat.* 8 (1968), 263–281.

CHAMOVE, A., H. F. HARLOW, and C. D. MITCHELL. "Sex Differences in the Infant-Directed Behavior of Preadolescent Rhesus Monkeys." *Child Development* 38 (1967), 329–335.

DEVORE, I. "Mother-Infant Relations in Free-ranging Baboons." In H. L. Rheingold (ed.), *Maternal Behavior in Mammals* (New York: Wiley, 1963), 305–335.

GARTLAN, J. S. "Sexual and Maternal Behavior of the Vervet Monkey, *Cercopithecus aethiops.*" *Journal of Reproductive Fertility,* Supplement 6 (1969), 137–150.

——— and C. K. BRAIN. "Ecology and Social Variability in *Cercopithecus aethiops* and *C. mitis.*" In P. Jay (ed.), *Primates: Studies in Adaptation and Variability* (New York: Holt, Rinehart and Winston, 1968), 253–292.

GOY, R. W. "Organizing Effects of Androgen on the Behavior of Rhesus Monkeys." In R. P. Michael (ed.), *Endocrinology and Human Behaviour* (London: Oxford University Press, 1969), 12–31.

HAMBURG, D. A. "Observations of Mother-Infant Interactions in Primate Field Studies." In B. M. Foss (ed.), *Determinants of Infant Behaviour IV* (London: Methuen, 1969), 3–14.

——— and D. T. LUNDE. "Sex Hormones in the Development of Sex Differences in Human Behavior." In R. Maccoby, *The Development of Sex Differences* (Stanford, Calif.: Stanford University Press, 1966), 1–24.

HARLOW, H. F. and M. K. HARLOW. "The Affectional Systems." In A. M. Schrier, H. F. Harlow, and F. Stollnitz (eds.), *Behavior of Nonhuman Primates,* Vol. II (New York: Academic Press, 1965), 287–334.

——— *et al.* "Maternal Behavior of Rhesus Monkeys Deprived of Mothering and Peer Associations in Infancy." *Proceedings of the American Philosophical Society* 110 (1966), 58–66.

HINDE, R. A., ROWELL, T. E., and Y. SPENCER-BOOTH. "Behavior of Socially Living Rhesus Monkeys in Their First Six Months." *Proc. Zool. Soc. Lond.* 143 (1964), 609–649.

——— and Y. SPENCER-BOOTH. "The Effect of Social Companions on Mother-Infant Relations in Rhesus Monkeys." In D. Morris (ed.), *Primate Ethology* (Chicago: Aldine, 1967), 267–286.

ITANI, J. "Paternal Care in the Wild Japanese Monkey, *Macaca fuscata fuscata.*" *Primates* 2 (1959), 61–93.

JAY, P. "Aspects of Maternal Behavior Among Langurs." *Ann. N.Y. Acad. Sci.* 102 (1962), 468–476.

———. "Mother-Infant Relations in Langurs." In H. L. Rheingold (ed.), *Maternal Behavior in Mammals* (New York: Wiley, 1963), 282–304.

JENSEN, G. D., R. A. BOBBITT, and B. N. GORDON. "Sex Differences in the Development of Independence of Infant Monkeys." *Behavior* 30 (1968), 1–14.

KAUFMANN, J. H. "Behavior of Infant Rhesus Monkeys and Their Mothers in a Free-ranging Band." *Zoologica* 51 (1966), 17–29.

———. "Social Relations of Adult Males in a Free-ranging Band of Rhesus Monkeys." In S. A. Altmann (ed.), *Social Communication Among Primates* (Chicago: University of Chicago Press, 1967), 73–98.

KAWAI, M. "On the Rank System in a Natural Group of Japanese Monkeys." *Primates* 1 (1958), 111–148.

KAWAMURA, S. "Matriarchal Social Ranks in the Minoo-B Troop: A Study of the Rank System of Japanese Monkeys." *Primates* 1 (1958), 149–156.

KOFORD, C. B. "Rank of Mothers and Sons in Bands of Rhesus Monkeys." *Science* 141 (1963), 356–357.

KUMMER, H. *Social Organization of Hamadryas Baboons: A Field Study* (Chicago: University of Chicago Press, 1968).

LAHIRI, R. K. and C. H. SOUTHWICK. "Parental Care in *Macaca sylvana.*" *Folia Primat.* 4 (1966), 257–264.

LOIZOS, C. "Play Behavior in Higher Prımates: A Review." In D. Morris (ed.), *Primate Ethology* (Chicago: Aldine, 1967), 176–218.

MITCHELL, G. D. "Paternalistic Behavior in Primates." *Psychology Bulletin* 71 (1969), 399–417.

———. "Attachment Differences in Male and Female Infant Monkeys." *Child Development* 39 (1968), 611–620.

——— *et al.* "Long-term Effects of Multiparous and Primiparous Monkey Mother Rearing." *Child Development* 37 (1966), 781–791.

POIRIER, F. E. "The Nilgiri Langur (*Presbytis johnii*) Mother-Infant Dyad." *Primates* 9 (1968), 45–68.

ROWELL, T. E., R. A. HINDE, and Y. SPENCER-BOOTH. " 'Aunt'-Infant Interaction in Captive Rhesus Monkeys." *Animal Behavior* 12 (1964), 219–226.

SADE, D. S. "Some Aspects of Parent-Offspring and Sibling Relations in a Group of Rhesus Monkeys, with a Discussion of Grooming." *American Journal of Physical Anthropology* 23 (1965), 1–17.

———. "Determinants of Dominance in a Group of Free-ranging Rhesus." In S. A. Altmann (ed.), *Social Communication Among Primates* (Chicago: University of Chicago Press, 1967), 99–114.

SEAY, B. "Maternal Behavior in Primiparous and Multiparous Rhesus Monkeys." *Folia Primat.* 4 (1966), 146–169.

SPENCER-BOOTH, Y. "The Behavior of Group Companions Towards Rhesus Monkey Infants." *Animal Behavior* 16 (1968), 541–557.

STRUHSAKER, T. T. "Auditory Communication Among Vervet Monkeys (*Cercopithecus aethiops*)." In S. A. Altmann (ed.), *Social Communication Among Primates* (Chicago: University of Chicago Press, 1967a), 281–325.

———. "Behavior of Vervet Monkeys, *Cercopithecus aethiops.*" *Univ. of Calif. Publ. Zool.* 82 (1967b), 1–74.

SUGIYAMA, Y. "Behavioral Development and Social Structure in Two Troops of Hanuman Langurs (*Presbytis entellus*)." *Primates* 6 (1965), 213–248.

VAN LAWICK-GOODALL, J. "The Behavior of Free-living Chimpanzees in the Gombe Stream Reserve." *Animal Behavior Monograph Series* 1 (1968), 161–311.

WELKER, W. I. "An Analysis of Exploratory and Play Behavior in Animals." In D. W. Fiske and S. R. Maddi (eds.), *Functions of Varied Experience* (Homewood, Ill.: Dorsey Press, 1961), 175–226.

Timothy W. Ransom and Thelma E. Rowell

Early Social Development of Feral Baboons

In this article we present some ideas that derive for the most part from Ransom's eighteen-month intensive study (2,555 observation hours) of feral baboon groups in the Gombe Stream National Park, Tanzania (1971; Ransom and Ransom, 1971). In confirming and extending these ideas we used the less intense, though longer, field study (twenty-two months; 380 observation hours) of baboons at Ishasha in the Queen Elizabeth National Park in Uganda by Rowell (1966, 1969) and a six-year study of a caged group of baboons in Kampala, Uganda (Rowell, 1968; Rowell *et al.*, 1968). These studies will be identified hereafter by the name of the place in which they were made. Previously, baboon infant development has been described from field studies in Nairobi National Park, Kenya, by DeVore (1963), and some data are also available on South African baboons (Hall and DeVore, 1965). The baboon thus occupies a unique position as the only species (or species group) of monkeys whose social behavior, especially infant development, has been extensively studied in the wild and in a variety of habitats. The majority of studies of infant development in primates has been made on captive or artificially fed macaques, and it is not always clear how developmental processes described in such circumstances relate to the organization of groups in habitats for which the organization presumably is adapted. In the two studies of feral baboon groups discussed in this article, the observers made no attempt to alter the natural behavior of the animals by feeding, marking, or any other form of intervention. The presence of the chimpanzee feeding station in the Gombe seriously affected the activity patterns of one troop for several months of the study period, but observations on this group have been used, for the most part, to supplement and confirm the main study of another troop. On the other hand, we do not yet have available for the baboon the type of detailed, intensive studies that have been carried out on captive macaques. Since comparisons made so far (Rowell *et al.*, 1968) suggest that macaques and baboons are fairly similar, at least in overall rate and general pattern of development, we have felt justified in making direct comparisons between the genera where needed to fill this gap.

DIFFERENCES IN THE ENVIRONMENTS OF INFANT BABOONS

Our findings differed in some respects from each other's, and from those of DeVore (1963), the other field study in which infant development has been studied in detail. Some obvious groups of factors, including differences in the physical and social environments, and in observation techniques, may have influenced the recorded pattern of infant development. Although we are not yet at the stage where we can assess the effects of these factors and their interrelations on this pattern, it is perhaps useful to set them out for comparison.

PHYSICAL ENVIRONMENTS

Both authors' studies were of forest-living troops, whereas DeVore's was of baboons inhabiting unusually open grassland. The forest, or forest-edge, environment is undoubtedly richer in food and water, less rigorously seasonal, and provides more cover than the drier savanna or wooded savanna areas. The immediate result of these environmental differences is that baboons do not have to go as far to feed in the forest as do DeVore's baboons. Open-country troops ranged up to ten or twelve miles a day (Hall and DeVore, 1965), and even their average trek of three miles was two to three times longer than that of forest baboons. Yet it was apparent that even after a comparably easy day, young animals at Ishasha were tired by nightfall, and did not play as they had earlier in the day. Due to the richness of individual food sources in the forest, foraging takes less time than in the savanna, and so there is much more time available for social interaction, or just sitting. It is probable that the frequency of interaction is itself important in determining social structure (Rowell, 1967). In addition, food from the forest tends to be softer and more easily available—compare tree fruits and the soft grasses of the forest forb layer with the dry grass storage leaf bases that form the staple diet of the savanna baboon in the dry season. Thus the infant forest baboon can feed himself more, and earlier, and so there is a different time relationship between his nutritional and psychological dependencies on his mother. On the savanna water is often scarce, and the few existing waterholes are shared by numerous other species, including predators. One of our studies was in a gallery forest along a permanent river, and the other was on the shore of Lake Tanganyika. Thus in both cases water was abundant and rarely associated with danger. This difference must be important in determining some variations in the nutritional aspects of the mother-infant relationship, as the forest-living female has readily available the great quantity of water required for milk production, while at the same time the infant is not dependent on her milk for water. Thus the infant, especially when older, may need to suckle less, and the transition from a physical to a mainly psychological dependence on its mother may occur sooner.

SOCIAL ENVIRONMENTS

In DeVore's study groups there was a clear though complex hierarchy expressed in frequent conflicts between males, while females were described as having a subtle and changeable hierarchy. At the Gombe Stream there were also discernible hierarchies in both sexes, but none was detected in either sex at Ishasha. In Nairobi the adult animals remained in the group throughout the study, whereas at Ishasha adult males moved frequently between troops. In the Gombe both sexes exchanged troops, but less frequently than the males at Ishasha. Moreover, the number of adult males and females was approximately equal at Ishasha, but females outnumbered males two to one in the Gombe and more than two to one in Nairobi. Since long-term relationships between adults, and especially between mothers and adult males, are shown below to have important consequences for infant development, these varying patterns of adult social organization must set different limits to the range of developmental patterns seen in troops of the three study locations.

The infant's social environment also depends on the distribution of births. In a group with a somewhat restricted birth season, as in Nairobi (October–December), the majority of infants is born into a cohort of close peers, with a long age gap between preceding and succeeding cohorts. In neither forest group was there a discernible birth season. Infants usually had playmates of all ages available, but only one or two close in age. These differences would be expected to affect both the time spent in play and the type of relationships established between infants. The birth interval was about two years at Nairobi, about fifteen months at Ishasha, and eighteen to twenty months in the Gombe. Due to the resultant differences in age gaps, relationships between siblings would be expected to vary in the three areas. Variations in the lactation interval before resumption of swelling (six months at Ishasha, ten months at the Gombe, and fifteen months in Nairobi) mean that after the first few months mother-infant interaction would take place with quite different backgrounds of maternal endocrine changes, and associated effects on the mother's social behavior. Birth interval and infant mortality combine to produce the adult female–juvenile ratio (i.e., the number of adult females compared to the number of juveniles), and both of these factors seem to be rather closely related to richness of habitat (Hall, 1963). The number of semidependent infants a female has in her "family" will affect relationships within it. A population with high infant mortality might not demonstrate to the observer the importance of sibling relationships at all, while the presence in such a group of adult females without dependent offspring could produce some of the behavior seen in newly established breeding groups in captivity, such as frequent infant-stealing.

OBSERVATION TECHNIQUES

We must explain the differences in observation methods at Ishasha and Gombe Stream, as these undoubtedly led to differences in emphasis and interpretation. Ransom's study is by far the more detailed, and data from it form the basis of the following sections. It was a continuous daily record of behavior over twenty-one months with the baboons being observed at very close range and known individually, so that individual relationships were stressed. Rowell's study took the form of repeated samples of behavior, observations being made for two weeks every two months for the first two years, and at greater intervals over the next three years. Fewer individuals were recognized, and behavior was referred to age-sex classes in general, rather than to individual relationships. Ishasha data were used in interpreting observations of mothers and infants in the Kampala caged group also studied by Rowell. Comparison of the two sets of field data allowed us to examine the generality of conclusions drawn from the Gombe Stream data, and to make some quantitative comparisons.

TABLE 1. Infants Observed at Ishasha, Gombe Stream, and Kampala

PLACE	NUMBER OF INFANTS	SEX	AGE	TYPE OF OBSERVATION
Gombe Stream				
Troop of 60	2	male	0–6 months	continuous, intensive
	2	female	0–6 months	continuous, intensive
	2	male	6–18 months	continuous, intensive
	1	female	6–18 months	continuous, intensive
	4	male	0–18 months	general, intermittent
		female	0–18 months	general, intermittent
Troop of 75	4	male	0–18 months	general, intermittent
	2	female	0–18 months	general, intermittent
Ishasha	65	male and female	birth, to birth of next sibling	general, intermittent
Kampala	6	male	0–3 months	intensive, regular
			0–4 months	intensive, regular

NOTE: In addition, these and other infants were subject to general observation for up to two years.

Table 1 summarizes the infant data on which our observations are based. Details of methods of recording behavior can be found in the three studies cited in the introductory paragraph of this article.

DEVELOPMENTAL STAGES

Table 2 gives a rough timetable for commonly used developmental stages in the olive baboon. Behavioral changes typical of each stage are listed below. Changes in patterns of social interaction are described in later sections. Although the lists are similar, there are some important differences between our timetable and that given by DeVore (1963). In general, we found that new behavior developed earlier than described by DeVore, although the mother-infant relationship continued into adult life in modified form and was by no means replaced by peer relationships as he suggested. For changes in time spent in various spatial relationships with the mother, see Rowell *et al.* (1968).

TABLE 2. Major Stages in the Development of Olive Baboons

STAGE	AGE	PHYSICAL CHARACTERISTICS
Newborn	1–4 weeks	naked skin bright pink; fur black
Infant 1	4–12 weeks	naked skin pale pink
Intermediate	12–24 weeks	coat turns to yellow or agouti, skin to dark gray (mother starts to cycle in Uganda)
Infant 2	6–12 months	coat yellow-gray and smooth; skin dark gray (mother pregnant in Uganda, starts to cycle at Gombe Stream)
Small juvenile	Second year	coat becomes grayer and shaggier (mother gives birth to new infant early [Uganda] or late [Gombe] in this period)
Medium juvenile	Second to third year	
Large juvenile	Third year: both sexes Fourth year: males	
Adolescence		
Female	End of third year	menarch
Male	End of fourth year	growth spurt; fertile
Maturity		
Female	Fourth to fifth	first infant; still looks young, can be confused with adolescents; stops growing at about six years
Subadult male	Four to six years	appearance and growth of canines and mane
Young adult male	About six to eight years	canines fully erupted

NOTE: Age estimates beyond two years are based on the Kampala study.

Adulthood: Estimates based on tooth wear of baboons from dry savanna woodland suggest that the life expectancy of both sexes is less than twenty years (Bramblett, 1967). Males have lived to be over forty years old in captivity. We feel that life expectancy in a forest habitat probably falls between these two figures.

NEWBORN

Day 1. Birth was not observed, but the infants were watched in the first few hours while still wet and occasionally with the umbilical cord still attached and soft; the mother may spend much of the first day sucking and chewing at the cord, and it was always gone by the next day. On the first day the infant seems unable to cling for long periods, and is frequently supported and adjusted by the mother; it may be off the nipple as much as half of the time. The mother avoids other baboons, feeds heavily, rests a good deal, and may place the infant on the ground occasionally.

Days 2 to 3. The infant clings to the mother better and is on the nipple about 80 percent of the time. It geckers and moan calls (immature distress vocalizations denoting anxiety, fear) in response to discomfort and may grin (fear face). The mother joins the maternal subgroup. She continues to adjust and support the infant during the first week, using elbows (when walking), knees, and hips (when standing bipedally, sitting, or climbing) as well as her hands. She adjusts the infant in response to its geckers or body jerks indicating discomfort.

Days 4 to 7. The infant is now able to cling tightly and unsupported for extended periods, and the mother has to support it for only the first step or two when she starts to walk. The infant starts to orient itself to the outside world, following movements with its eyes, and turning toward sounds, letting go of the nipple to do so. It may respond to the lip smacking of investigating adults by giving a play face (wide-open mouth) and waving a hand toward them. By the end of the first week the infant starts to turn around in its mother's lap, and can reach her nipples and shift from one to the other without assistance. Riding posture is almost exclusively ventral, but occasionally an infant will ride across its mother's back, supported by her tail.

Week 2. The infant first leaves the mother voluntarily, and may remain on the ground beside her for ten to fifteen minutes at a time. When taken from her by others, the infant is able to crawl back to the mother with legs well flexed. The infant is able to stand bipedally with support, but the hind legs are weak. It begins to reach for and mouth grasses in a poorly coordinated manner, but does not yet put objects to its mouth with its hand. Play between mothers and infants was first seen on Day 9. Crouching and lip smacking (invitation to approach) at the infant by others from a distance was seen on Day 10.

Week 3. The infant starts to totter away from its mother but is usually retrieved fairly quickly. Interactions with others increase, but bouts are short and alternate with long periods of nursing and sleep.

Week 4. The infant begins to initiate regular contact with other infants and adults. It drops to the ground as soon as the mother stops walking now. Frequent mouthing and chewing of the mother, especially her knees, by the infant may be associated with teething.

INFANT 1

In general, this period sees the major change in mother-infant spatial relations—by the end of it the infant is frequently away from her for distances up to 20 feet, and occasionally up to 200 feet. In the second month, climbing on the mother in exploration develops into dorsal riding. The infant pays increasing attention to the environment, handling and mouthing objects—including mother's food, which gives rise to some conflict between them. Soft food is eaten in small amounts. Male infants start to show the penis flick in response to grooming. Threat (eyes wide, head jerk) was seen in the eighth week. In the third month, the infant begins to spend half his riding time upright on its mother's back, at first with support from her raised tail, and then later unaided. Time spent in exploration and play continues to increase, and the infant begins to play in trees as well as on the ground. When the mother is foraging, the infant will follow her more and more rather than riding. Pelvic thrusts against an adult male were seen in Week 10. In the fourth month, following rather than riding becomes the usual pattern during foraging. By the end of the period most of the infant's waking time is spent in play with other infants and juveniles, and sexually dimorphic play patterns become apparent. More adult social behavior emerges: Response to presenting of a swollen female by mounting and thrusting (badly oriented) was seen in Week 14, and masturbation to erection in Week 15. An approximation of the "double bark" (a vocalization emitted by older immatures and adults in a variety of contexts, including separations from the rest of the troop) was heard in Week 15.

TRANSITIONAL PERIOD

In the fifth and sixth month the coat completes its change from the natal black to a golden brown. The infant continues to extend the time and distance away from its mother and is now frequently more than 200 feet away for ten or twenty minutes at a time. The infant is often out of sight, and out of reach of effective maternal assistance. The range of food taken increases, and includes hard fruit by the end of the fifth month. Food-cleaning behavior was first seen in Week 17. Play with objects, peers, and juveniles increases still further, and new communication patterns develop: The foot-back present develops from genital investigation by adults. More threat patterns, including stamping, slapping branches, and displaying eyelids, were seen in Week 18, and pant grunting (a medium intensity threat vocalization) in Week 20. The infant now grooms adults. Sexual behavior becomes more obvious: Males respond to presenting adult females by placing hands on hips; males and females mount in play; males approach and mount female juveniles. The sexes begin to separate in play groups, and females occasionally carry small infants. Tolerance by adults diminishes, and they attacked the infants in Week 21. At Ishasha, the mother may show first post-parturition estrus during this period.

INFANT 2

The infant is away from its mother for most of the day, one to two hours at a time, and frequency of interaction with her declines to a low point just before the birth of the next sibling. Weaning behavior is seen most frequently in this period, and the infant forages with the mother, eating what she eats. Female infants may move about with individual young adult males, or remain with mother-infant groups, and males begin to play mainly with older juvenile males. Gombe Stream mothers showed first post-parturition estrus in this period.

SMALL JUVENILE

In the second year males join a relatively permanent peer group, and females mostly avoid rough play and spend much time with new mothers. Males now show a complete copulatory pattern. By the second half of this period the mother usually has a new infant, and there is an increase in interaction with the mother. The juvenile is sometimes carried by the mother, dorsally and occasionally even ventrally, especially when she is swollen. Elder siblings of this age interact with the new infant extensively (especially female juveniles). They groom the mother and are groomed by her, and sleep with her until after the end of the second year. The mother rarely supports her juvenile in squabbles, however, and competes for food with him without consideration.

THE RELATIONSHIP BETWEEN INFANT AND MOTHER

The bond between mother and infant is probably the most intense and long-lasting relationship that a baboon may experience. Its dependent nature demands that the participants arrive at a mutually satisfying pattern of interaction whose consistency and flexibility allow the physical and social maturation of both to follow their proper courses. The interaction pattern depends for its original form and later alterations on the characteristics both of the pair itself and of the social and physical environment. Thus the age and maternal experience of the mother affect her behavior from the very beginning, including her degree of "restrictiveness" or "permissiveness," as well as the success with which she satisfies her own and her infant's needs with ease and economy of effort. Her rank or status in the group, itself partly dependent on age and experience, determines how others in the group will relate to her and the infant, which in turn appears to affect both the group's and the infant's expectations of the latter's future rank and relationships. Koford (1963) found a similar correlation in rhesus monkey groups. At the same time the sex and degree of physical maturation of the infant—its developing ability to move, interact with others, and sustain itself for increasing periods of time—produce progressive changes in the interaction pattern. The composition and the stability of the group can affect the relationship: The availability of peers and playmates, the presence or absence of siblings, and the amount of tension or conflict within the group (determined by the adult male/female ratio, the availability of food, the threat of predation, and so on) determine the limits of the form and manner of development of the relationship. Finally, the particular genetics and experience of the mother-infant pair combine with these general factors to produce the individual patterns of interaction observed.

When her infant is born a female baboon has the problem of integrating those behavior patterns necessary to support the infant's development while at the same time maintaining her own role in the group and providing for her own sustenance and protection. She must achieve a compromise, a set of behavior and interaction patterns that best satisfy both her own needs and those of the infant and that keep conflict between these needs to a minimum.

The pattern of change in the basic contact and nursing interactions of baboon mother-infant pairs over the first three months in a caged group situation have been described by Rowell *et al.* (1968). The development of the infant baboon was found to be very similar to that of the rhesus monkey infants reported by Hinde *et al.* (1964) under equivalent conditions, especially in terms of changes in mother-infant spatial relations. The baboon infants matured slightly faster than the rhesus, but the difference might have been due to the very different climatic environments of the two colonies. Observations of mother-infant pairs in wild groups at Gombe Stream and Ishasha confirmed the rate of infant development found in captivity, so that it is possible to make direct comparisons between these studies.

In both species, and in both cage and wild environments, certain clusters of behavior stand out from the continuum of mother-infant interaction because they involve conflict between the two. They attract the observer's attention just because for the most part the two animals are so well synchronized that their smooth interaction can be taken for granted. The results of this harmonious relationship are discussed in the preceding section; some types of conflicting behavior are discussed below.

RESTRICTION

In caged groups both baboon and macaque mothers frequently and actively restrained their infants during the first few months. The behavior included restraining the infant in its attempts to move out of contact, hovering close when it was out of contact, and frequent retrieving, and was seen most often during the first two to three months of the infant's life. After the third month it became less frequent, and macaque mothers did not restrict after the twelfth month. Restrictive behavior was associated with the mother's avoidance of other adults, and often with frequent attention by high-ranking females to the infant. Restrictive behavior by baboon mothers effectively reduced the time the infants spent away from them, but did not seem to be especially effective in its apparent intention of limiting interaction with other adults in the colony.

Mothers have been observed to restrict their infants fairly frequently in free-ranging rhesus troops (Kaufmann, 1966), as well as in the caged situation, so it is perhaps a normal component of the mother-infant relationship in this species. This is not the case, however, for baboons. In the wild groups observed at Nairobi, Ishasha, and Gombe Stream mothers were rarely seen to restrict an infant's movements, in contrast to the mothers in the caged group, all of whom, except the highest ranking, restricted their infants to some extent during the first three months. In two extreme cases, mothers whose infants were of special interest to the high-ranking female would spend up to an entire observation hour fighting their struggling infants. DeVore makes no mention of restrictive behavior in his description of mothers and infants in Nairobi. At Ishasha and the Gombe restriction was seen very infrequently. On rare occasions a mother was seen to prevent an infant from moving farther off by grabbing a leg or tail, usually in response to a sudden disturbance, such as the appearance of the observer or the approach of fighting males. Certain other behavior of mothers toward their infants seemed designed to reduce their movement, but could not be called restrictive in the sense of preventing independent activity or interaction with others. For example, in the first week or two most mothers were seen to pin their infants against their chests with knees or thighs (Figure 1). This posture seems supportive, rather than restrictive, in that the infant is at first unable to support himself against the chest and maintain nipple contact for prolonged periods. Mothers of older infants were also occasionally seen to grab an infant by the hair or foot when it tried to withdraw from her grooming. As the mother's grooming was most often an attempt to divert the infant's attention away from the nipple (see below) this behavior was hardly restrictive in the sense used above. Some wild mothers hovered close and retrieved fairly commonly, but only in a few instances did this seem to conflict with the infant. Infants were never seen to actively avoid their mothers (rather, the reverse was usually the case), and a negative response to retrieving, such as refusing to cling, was rare. Most often infants would accept being retrieved and then go off again at the first opportunity, or if being carried, release its grasp with arms or legs and drag on the ground.

Restriction, however, was not an unusual feature of an infant baboon's experience. In the Gombe troops infants were fairly often picked up or taken off the mother by other adults, especially males, or by two-to-three-year-old juveniles. They would carry and groom the infants and restrain them if they tried to move away (Figure 2). (See section on "Relationships with Siblings and Other Juveniles.") It appears, however, that frequent and intense conflict between mother and infant due to the former's restraining and retrieving behavior is a cage artifact. In the Kampala group it seemed to be related to the existence of a rigid hierarchy among the females and the persistent attempt of two high-ranking females to steal some infants. The degree of restrictiveness was not, however, correlated with rank, and in the wild troops of Nairobi and Gombe Stream there were female hierarchies but very litle restriction. Rather, it appeared to be related to the caged female's inability to control her social environment. In wild groups, others paid a great deal of attention to the infants, especially during the first few weeks, but the mother could avoid the majority of these attentions in a number of ways. She might simply move or turn away, in which case the other usually became involved in some other interaction, or in feeding. If the attentions were persistent, the mother might briefly attack, or else move closer to a friendly adult male or a high-ranking female

Figure 1

Figure 2

Figure 3

whose presence supported an attack or inhibited the unwanted attentions. Which method a female used depended on the age, rank, and experience of the participants, and on immediate circumstances, but one or more such recourses are almost always open to her. In the cage, with limited space, small group size, and abnormal group composition (for example, only one adult male present), the possibility of avoiding interactions was much less, while tension was high and aggression relatively frequent. The resultant protectiveness, which led to conflict between some mothers and their offspring, was reduced immediately by temporary, partial isolation of the mother-infant pair (Rowell, 1968).

"REJECTION"

Facilitating Early Independence. Rather than restrict her infant in the first few weeks a wild baboon mother may actively encourage its independent activity. In the Gombe troops many females placed their infants on the ground beside or in front of them on the first day, and continued to do so frequently thereafter. One young mother, Faith, put her infant Charity on the ground, and on some occasions walked a few steps away from her, thirteen times in four hours of observation in the first five days (Figure 3). These early separations were usually ended by the approach of a juvenile or other adult, when Faith would retrieve her infant. By the end of the first week Charity was taking her first wobbly steps and on some occasions reestablished contact with her mother herself. It is notable, considering the effects of "baby stealing" in the cage group, that Faith continued to encourage Charity's independent movement in this way even though another female stole and kept the infant for over three hours on her third day.

Infants were treated in this way at different ages by Gombe mothers, and the differences seemed to be correlated with the age, rank, and reproductive history of the mothers. A female baboon's nipples are stretched and become elongated and probably somewhat desensitized by the sucking and pulling of her infant. They may shrink again following prolonged disuse, but among multiparous mothers the elder offsprings' continued attention to the nipple almost up to the birth of the next infant maintained the nipple's length of two to three inches, while the nipples of primiparous mothers may be only one-half inch long. During the first few days the infant needs assistance from its mother in finding and maintaining contact with the nipple. Its ability to cling in a tightly flexed position against her upper chest, whether the mother is sitting or walking, is limited, and the mother frequently has to support and adjust her infant in the early period. When she sits for long periods, grooming or resting, the infant will relax its grasp and sink down into her lap or onto her feet. A first-born infant will not be able to maintain contact with its mother's short nipple, and will soon begin to struggle and search for it again. An infant may be able to retain the longer nipple of an older mother and will often go to sleep in the slumped position (Figure 4). This small difference in nipple length may produce extensive variation in the early experience of infants, and in the development of mother-infant relationships. During the first few days young females have to interrupt their own feeding and social activities frequently to assist their first-born infants to reach or maintain contact with the nipple. At the same time they are probably experiencing discomfort from the stretching of the nipple and set their infants down fairly often. Their infants in turn experience frequent frustration when on their mothers, and experience loss of contact and support from the first day. The early interaction between multiparous mothers and their infants appears much more harmonious and secure. These infants are able to keep the nipple when the mothers are seated or walking without having to maintain tight flexion, and thus make fewer demands for attention and support. The mothers seem to be less disturbed by the infants' manipulation of the nipple, and are less likely to put their infants down.

Length of nipple is one aspect of what has been called maternal experience. Variations in early experience due to this and other factors may have far-reaching effects on the development of the mother-infant relationship, and on the infants' social development. In their study of rhesus monkeys, Hinde and Spencer-Booth (1967) found considerable stability over time in the rank order of mother-infant pairs on a number of measures descriptive of their relationship, such as the time spent on and off the mother, the proportion of leavings and approaches initiated by each, and the frequency of maternal rejection. The relationship established in the first few days may thus become the basis of the pattern of interaction throughout the duration of their close association. If so, it is possible to analyze later variation in mother-infant interaction in terms of the earliest experiences of the pair. For example, during the second and third months some multiparous mothers began to encourage greater independence in their infants, making them follow rather than being retrieved or riding, by suddenly and quickly moving away from their infants who had moved off a few steps to play or explore.

The following rather extreme example is from the Gombe Stream:

Myrna (a mature multiparous female) approaches the stream with her twelve-week-old son Moley riding on her chest. She sits a few feet from the stream, which is six feet wide at this point and running heavily today. Moley moves off a few steps to pick at some leaves. Suddenly Myrna jumps across the stream, leaving Moley on the other side. Moley begins to scream as his mother stops on the other bank and looks back at him. After about fifteen seconds he jumps in and is immediately carried several feet downstream. Finally he makes contact with the bottom, and his screams now gurgle, he struggles across to the opposite bank. As soon as he arrives Myrna jumps back to the opposite bank again, landing a little farther downstream. Moley immediately jumps back in, still screaming, and is again carried several feet closer to the lake before making it to the other side. Moley climbs up the bank and charges after his mother who has disappeared into the bushes.

More typical instances involved the mother simply running off through the group, so that the infant had to make his way past interested adults and juveniles to establish contact. One mother would play "tag" with her infant around an adult male, forcing it both to follow and interact extensively with the male. The mothers who behaved toward their two-to-three-month-old infants in this way were the same mature females who showed little tendency to initiate separation during the first few weeks. The infants of the mothers who had encouraged early separation were by this time already frequently following, rather than being carried by, their mothers and spending a good deal of time out of contact and interacting with other animals. Instances of forced following and sudden separation were rarely seen in these pairs. The frequency and intensity of such interactions between multiparous mothers and their infants suggested that the infants' expectations of contact and support, combined with their increased size and physical demands, were now conflicting sufficiently with the mothers' social and maintenance activities to require a significant change in their relationship, much to the distress of the infants. At this age then, first and later born infants interact differently with their mothers in ways that related to differences in previous interaction patterns. Conflicts of this type were possibly more common, and more extreme, in Gombe Stream than in other populations, so that they may be studied there in greater detail. DeVore did not describe such behavior from the Nairobi troops. Two primiparous mothers out of four in the caged Kampala group left their infants in the earliest weeks as described here, and a single instance, also involving a primiparous mother, was seen at Ishasha. Similarly, forced following by two-to-three-month-old infants was seen at Ishasha in only a few brief interactions.

We have seen that the infant appears to be rather conservative, the mother innovative in the initiation of independent behavior. Similarly, Hinde and Spencer-Booth (1967) found that the mother rhesus monkey plays the major role in facilitating independence of her infant in the early months. The infant baboon must be sufficiently independent by the time the mother ends her lactation interval to allow her the necessary time and energy for estrous behavior. Changes in the relationship do not take

Figure 4

Figure 5

Figure 6

place suddenly, nor are they entirely dependent on the mother's initiative. They occur only when the infant has matured to the point where his own behavior and the resultant interaction pattern of the dyad offer an alternative form for the relationship, as the following example illustrates.

Methods of Carrying. Beginning about the seventh or eighth week infant baboons start to ride on their mother's back (dorsal) rather than exclusively on her chest (ventral). This change appears to be the natural result of the infant's increased ability to support himself in a variety of postures, the reduced amount of time he needs to spend ventral on the nipple (because of more efficient sucking and increased confidence), and his curiosity. Among mother-infant pairs in the Gombe Stream, this riding posture resulted on the first few occasions from the actions of both mother and infant. At this age, while the mother remains stationary, an infant spends much of his time exploring the immediate environment and in the process frequently climbs on his mother (Figure 5). As early as the sixth week the mothers began to stand up and walk off while their infants were climbing on them. At first the infants would cling where they were for a few seconds and then slide down to the ventral position, but later they would climb up and lie across her back. With practice, and on the infant's own initiative, this position soon evolved into the lying dorsal one (Figure 6) and eventually into the jockey position. By the tenth or twelfth week the infant may be spending half or more of its riding time on the back, but is quick to adopt the ventral position to contact the nipple or at the first sign of danger. Only when the infant has mastered the dorsal riding does it become an alternative to the ventral one. Then the mother begins to encourage the infant to ride dorsally more and more frequently—his greater weight is easier to carry on her back, he interferes less with her own movement, and her nipples are less accessible. At first she may do this by moving off a few steps and then, as the infant approaches, lowering and offering her back, which the infant will usually cling to or climb onto. The benefits of the dorsal riding posture to both mother and infant are especially apparent when the mother is foraging: Rather than having its play and exploration limited to the times when his mother is stationary, the infant is free to jump on and off his mobile mother for frequent, brief forays without disrupting her activity. Later on, however, mothers began to use various methods to discourage infants from riding at all, especially ventrally, rather than walking. For example, mothers with an infant on the chest were occasionally seen to stop and hunch their backs and shoulders, forcing the infant off onto the ground. If the infant was dorsal, the mother might reach back and slap, pinch, and poke him until he dismounted. When an infant persisted in his attempts to ride, the mother might eventually bite his wrist and throw him to the ground. A typical sequence of rejections by the multiparous Myrna of her thirteen-to-fifteen-week-old son Moley was as follows:

2/12/69 11:45 A.M. Myrna is feeding with others by the stream; Moley is walking beside her. An adult male, Moon, fights briefly with another mother for possession of her infant and then approaches Myrna, eyeing Moley. Myrna steps away from Moon with Moley ventral, then stops and pushes Moley off her chest to the ground. Moley climbs onto her back and sits up jockey dorsal; Myrna continues on.

2/19/69 1:10 P.M. Myrna forages with a small group along the edge of the stream; Moley rides ventral. Suddenly she grabs his arm and tries to pull him off her chest. Moley screams but retains his hold and slides lower onto his mother's stomach. Myrna struggles with her son for a few seconds and finally succeeds in pushing him off onto the ground. Moley screams and gecks, his hair standing on end, and begins to climb onto Myrna's back. She gives a slight jerk; he drops off and throws a brief tantrum, crouching on the ground, screaming and making small jerky movements with his entire body. Moley starts to climb onto his mother's back again; she hits back at him with a hand and knocks him off. Moley continues to scream and geck as he attempts to climb dorsal a third time. Myrna reaches back, grabs Moley by the arm and puts his hand into her mouth, pulling him off her back in the process. She walks forward three steps, carrying Moley only by the fingers in her mouth, and then drops him. Still screaming, Moley tries to climb up his mother's side. Myrna stops, pulls him off, and then goes on again; Moley follows screaming, his hair standing on end. He tries to climb up Myrna's leg; she shakes him off and then sits down, her knees pressed close together. Moley struggles for seconds to get in between them, then Myrna lets him into her lap as two men approach from the south.

1:13 P.M. Myrna starts to walk south again; Moley walks a few steps underneath her, gecking, and then climbs onto her back and lies down. Two minutes later he is off in play with another infant.

Weaning. At some point baboon mothers begin to reduce the infant's reliance on them for contact, food, and physical support. As we have seen, the baselines for duration of contact, suckling bouts, and so on, appeared to be set for each pair by the mother's behavior in the first few days. Some reduced the total time in contact or on the nipple by frequently setting the infant down, others apparently allowed their infants as much contact and time on the nipple as they wanted for a much longer period. There came a time when each mother began a second phase in contact reduction by actively rebuffing her infant in attempts to suckle or mouth the nipple. The time and form of this rejection, which can be called weaning, varied considerably among mothers at Gombe Stream. At the present state of

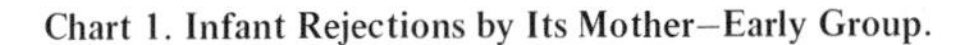
Chart 1. Infant Rejections by Its Mother–Early Group.

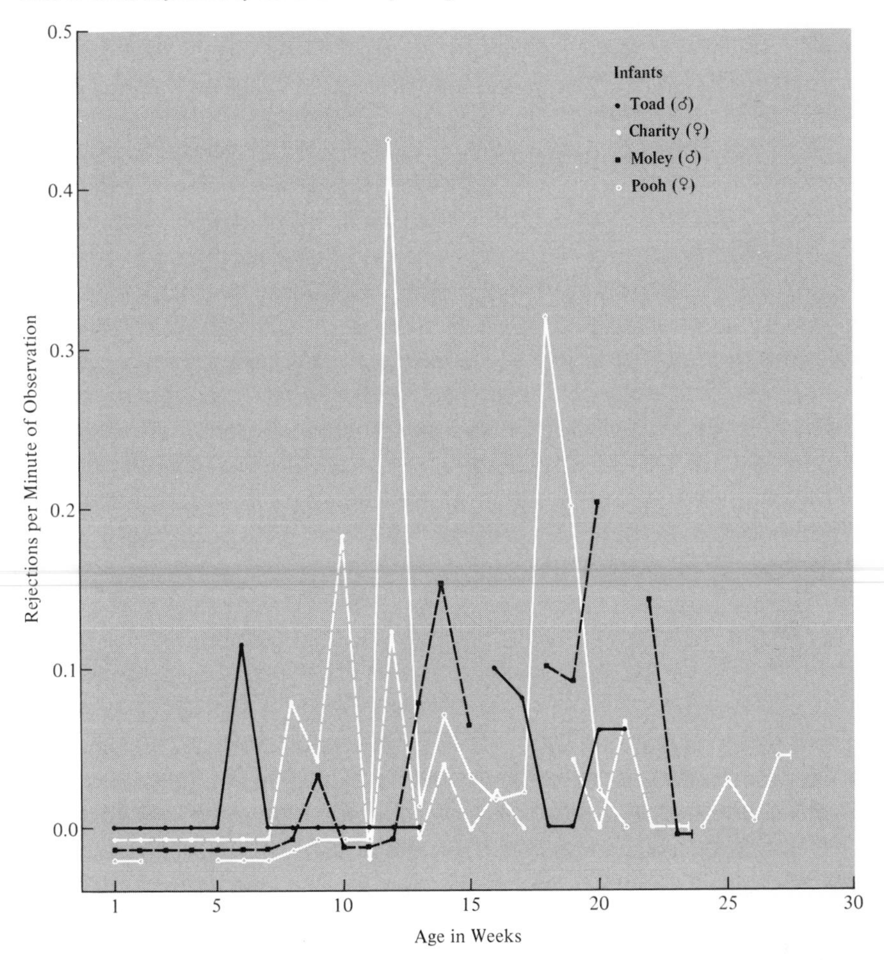

analysis, some of this variation can be related to parity, and it is likely that there will be further correlations with social environment. Some young mothers, for example, would often attempt to divert their infants' attention from the nipple by grooming or playing with them, methods that were not often successful and were only rarely seen in multiparous mothers. Other techniques included blocking access to the chest, either by subtle or natural movements of the hand like carrying food to the mouth or simply by repeatedly pushing the infant away. Myrna preceded the conflict with Moley over riding positions with the following interaction, which is typical of this type of behavior:

2/12/69 11:15 A.M. Moley has been off in play with juveniles for one min-

Chart 2. Infant Rejections by Its Mother–Late Group.

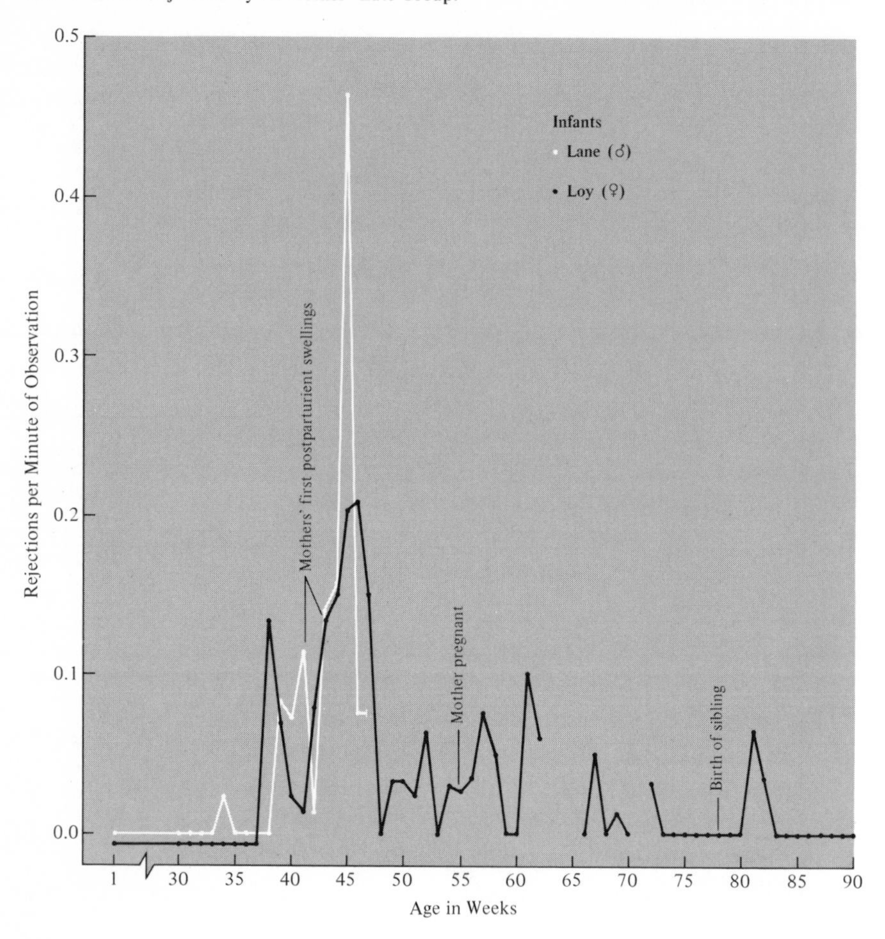

ute, fifteen seconds. He runs back to Myrna, who is seated and grooming herself, and starts to climb into her lap. Myrna lifts her arm quickly, bumping Moley and pushing him out of her lap. Moley leans forward again and takes the nipple in his lips for fifteen seconds, then runs off to play. Six minutes later he returns, climbs into his mother's lap, takes the nipple in his mouth again, and eventually goes to sleep.

Charts 1 and 2 illustrate the division of Gombe Stream mothers into two groups according to the age at which they began to reject their infants significantly often. Some mothers rejected repeatedly as early as Week 10 and continued to do so for at least the next ten to twenty weeks (early group, Chart 1), whereas others did not begin to reject until their infants were thirty-five to forty weeks old (late group, Chart 2). The later period coincided with the appearance of the mother's estrous cycle and continued into the next pregnancy.

It seemed that mothers who began to reject their infants early did not have to do so as severely during the later period as did mothers who waited. Perhaps because of the longer period of greater harmony and security that preceded rejection, the late-group infants showed greater persistence in their attempts to suckle than did infants that experienced earlier rejection, and they appeared to be much more disturbed by the mother's change in behavior. At the height of the weaning period these infants would often spend several hours a day fighting with their mothers to suckle, mouth, or hold the nipple in their hands, and frequently would throw prolonged tantrums.

DeVore describes rejection and weaning behavior in Nairobi groups, but did not see it until the ninth or tenth month—that is, he had no examples of early weaning. At Ishasha also, weaning was only seen in the second six months, and was never so extreme or prolonged as that seen in the Gombe. In a typical interaction, a female pushed her infant away once, just after a prolonged nursing bout, but he was allowed to take the nipple at his second attempt. Similar single rejections of the same infant were seen half a dozen times in a two-week observation period. There were fewer hours of observation at Ishasha than at Gombe Stream, however, so evaluation of the full extent and significance of these regional differences will require additional study.

As illustrated in Chart 2, the incidence of rejection and conflict between late-group mothers and infants rose markedly as the mothers began to cycle again. As DeVore suggests, this increase appears to be due to the sudden change in the mothers' behavior and the increased incompatibility of her activity, now including sexual and consort behavior, and the demands of the infant. By this time, however, the infant is eating most, if not all, available foods and has a fairly complete repertoire of social and locomotor behavior. The nature of its dependence is more psychological than physical at this age, and repeated and determined rejection by the mother during menstrual cycles and early pregnancy reduce the amount of contact between them to a minimum a month or so before the birth of the next infant. This minimum is still a fairly high level of interaction—the infant sleeps with its mother and is groomed far more by her than anyone else. The infant turns more and more to peers, older juveniles, and other adults for the contact and support in social interactions that its mother no longer supplies.

The importance of the resumption of cycling and later pregnancy in accelerating weaning is illustrated by the case of Icarus, whose mother began to reject him in the second six months, but then failed to resume cycling (she was very old). During the next twelve months the relationship between Icarus and his mother remained basically static: Although rejected fairly frequently, Icarus continued to remain close to his mother and to suckle and ride dorsal often. By the end of the second year he was still associating with his mother for most of the day and frequently taking the nipple (L. Taylor, p.c.[1]). On the other hand it will be remembered that mothers in Uganda began to cycle before any significant amount of rejection was noticed, and mothers would alternate bouts of sexual and maternal behavior without showing weaning behavior. The relationship between the two sets of behavior must vary according to the timing.

In describing the mother-infant relationships of baboons, DeVore suggested that during the second year the emotional ties between infant and mother are severed. By the end of this period the juvenile is no longer dependent on his mother, and seeks its social interactions and learning experiences in the group as a whole. Our observations, both in captivity and in the field, agree with those of other primates (rhesus macaques, Sade, 1965; chimpanzees, van Lawick-Goodall, 1968) that report the existence of complex and durable systems of interrelationships, based on familial and sibling ties, within the social group. Independent social activity is not dependent on, or the result of, the dissolution of the infant's bond with its mother. Rather the relationship undergoes a progressive series of changes that accommodate developments in the social and physiological status of both. Positive and preferential interaction between the pair continues into the offspring's adulthood, and there is a suggestion, though observation has rarely been of sufficient duration to prove the point, that many of the special relationships seen between adults, such as enduring preferences for sexual and social partners, may stem directly from familial relationships recognized by the maternal link. Some examples of the probable ontogeny of such adult relationships are given in the following sections.

RELATIONSHIPS WITH SIBLINGS AND OTHER JUVENILES

The relationship between mother and juvenile, a durable pattern of positive and preferential interaction, affected the structure and activity of the group and the social development of subsequent infants. A description of the interactions in one family, involving the mature female, Myrna, mentioned above, and her offspring, Loy and Moley, will illustrate the importance of such relationships. Loy experienced rejection and weaning by her mother in the late period (cf., Chart 2). In the month prior to Moley's birth she and Myrna made contact relatively infrequently during

[1] p.c.: personal communication.

the day, but groomed together almost exclusively and slept next to each other at night. In addition, they shared a number of relationships with other juveniles and adults, especially the adult male Harry, and as a result spent much of the day at least within sight of each other. With the birth of Moley, Loy intensified her attentions to her mother, often attempting to sit in contact or in her lap, grooming her frequently, and also touching and grooming Moley whenever possible. After about a week she also began to orient to the breast again, and in the next two months was observed to suckle and mouth the nipple three times. This regressive behavior elicited additional rejection from Myrna for a few weeks, and eventually Loy desisted in her attempts to attain close ventral or dorsal contact, and the nipple. However, during the next five months Loy continued to interact with her mother and brother more than anyone else, and at the end of the study was still spending a major part of her time with them.

The first animal other than his mother with whom Moley interacted extensively and positively was his sister. During the first two weeks Myrna continued to groom almost exclusively with Loy, and allowed her to nuzzle, lick, and groom Moley, liberties that no other juvenile could take to such a degree. As Moley began his first brief trips away from Myrna, Loy was often with him, shepherding him about and protecting him from the most energetic attentions of other juveniles. The following sequence involving Loy and the nine-day-old Moley illustrates their relationship:

11/27/68 9:37 A.M. Moley is on the nipple and Myrna grooms Loy. The adult male, Harry, and several other juveniles and females are close by.
1'23 [2] Myrna moves six feet away from Moley when he comes off the nipple, Loy follows and is groomed again. . . .
2'05 Moley climbs onto Harry's leg, giving distress moans and fear grins. Myrna moves closer to Moley, then off again, as the infant crawls toward her and Harry follows him. . . .
4'00 Loy has moved off ten feet. Moley starts to crawl away from Harry and toward his mother again, then turns toward Barbie and her three-year-old daughter, HW. Harry follows and sits close to the infant again. Moley continues to crawl toward HW who appears to be tense. HW backs away and then crouches at Moley. The infant crawls onto her foot, HW jumps away, Harry watches and grunts. Loy approaches, HW moves away from Moley who immediately screams. Moley climbs into Loy's lap as she sits right in front of him. HW glances at Harry and then sits behind Loy and reaches around her to touch Moley on the back. Moley continues to moan. Barbie approaches and presents to Moley and Loy, then sits behind. HW stands up and puts a

[2] This and subsequent numbers indicate the passage of time after the action commenced. For example, 1 minute 23 seconds after the grooming commenced, Myrna moved away from Loy.

hand on Loy's back. Loy moves away, picking Moley up. Moley starts to geck and struggle after a few steps, Loy sets him down.
6'35 . . . Moley stays with Loy. Myrna starts toward them but is intercepted by Harry. Moley crawls toward his mother, moaning and gecking, HW and Loy follow. HW reaches out to touch Moley, Loy puts an arm around him. Myrna walks over to Moley and briefly sniffs him, then moves off again, Harry follows her. Loy embraces Moley again, HW pushes and knocks her down, causing Moley to squeak. Loy chases HW off and then returns to groom Myrna who is now three feet from Moley. . . .
9'00 Moley crawls around Harry and back into contact with his mother. Loy stops grooming to embrace Myrna, then grooms again. Moley sits in front of them and moans. . . .
9'30 Loy stops grooming and reaches around Myrna to poke Moley who gecks . . . Harry starts to pull on Moley who has taken the nipple, then lets go, Moley suckles. Myrna grooms Harry. . . .
10'50 Loy returns, embraces Moley on Myrna's chest, then licks his ear. . . .
12'30 Myrna is napping. Loy sits in contact beside her and touches Moley who is still on the nipple. . . .
13'10 Loy grooms Myrna for ten seconds, then pokes her head through Myrna's elbow to look at Moley who is suckling now.
13'45 Loy sits with her back against Moley, Myrna is still napping. . . .

Loy's relationship with Myrna and Moley is an extreme example, and was not typical for all infants in the Gombe troops. Several other female yearlings in the same troop did not show a comparable increase in interactions with their mothers following the birth of their siblings, and no one else showed comparable regressive behavior. However, in every case in which the elder sibling was known with certainty, mothers continued to interact positively with their previous offspring, grooming them frequently and supporting them in most aggressive social interactions. Undoubtedly the special nature of Loy's relationship with her mother was due in part to the latter's high rank and the consequent protection that she offered, as well as to the strong relationships with others, especially Harry, that they shared. As will be described in the next section, Harry interacted frequently and positively with both of Myrna's offspring in their infancies, and continued to provide support and contact for Loy after weaning and Moley's birth. All infants were observed to interact extensively with available siblings, especially in the first six months, and these interactions may well be the basis for some of the social bonds found among adults.

In the Gombe troops some mature females shared intense relationships with three-year-old females, which, according to several lines of evidence including marked physical resemblance, appeared to be based on familial or sibling ties. The interaction patterns between these females and juveniles, both before and after the birth of the new infant, closely resembled those between known mother-infant pairs, and included frequent proximity, contact, and grooming. The three year olds were allowed to take extensive liberties with the new infants of their presumed mothers, but were punished more readily than younger siblings for overenthusiastic attentions or rough handling. In addition, on occasion each of these juveniles was observed to embrace the mother and infant, and in the process to place its mouth on the mother's chest close to the nipple. This exact form of the greeting pattern was not seen among any other combination of individuals.

A strong and persistent relationship including this unusual greeting pattern between a mother and her three-and-one-half-year-old nulliparous daughter was observed in the caged group as well. Two-thirds of the mother's interactions with the seven available females in the group were with her daughter, who in turn interacted with her own mother more than with any other. The daughter interacted with all infants frequently, but significantly more with her female sibling than with any other during Weeks 4 to 12. However, as in the Gombe troops, the mother's attitude toward her older daughter's interactions with the new infant was not noticeably more permissive than toward those of other females.

The finding of pervasive and persistent familial and sibling relationships among baboons supplements similar observations in other primate groups. Among chimpanzees, for example, the most obvious and stable social unit appears to be the family group, involving the mother and her successive offspring (van Lawick-Goodall, 1968). Sade (1965) has documented the long-range effects of familial relationships upon the structure and activities of the social group among the rhesus macaques of Cayo Santiago. Distinct relationships between female rhesus and their offspring are maintained into the latter's physical maturity and continue to affect patterns of social activity, including spacing, grooming, and the development of aggressive encounters within the group.

The social environment of the new infant may be highly structured by the set of relationships between its mother and her previous offspring. The attentions of these elder siblings to the young infant, and the contact and support they provide, especially when out of contact with the mother, frequently allow the latter to spend less time in the actual care of her infant than primiparous mothers, and to interact more extensively with others in the social group. At the same time the infant is experiencing frequent, usually positive interactions with its siblings.

An infant's interactions with its siblings account for only a portion of its total experience with other immatures and adults. A number of observers of primate groups have reported the intense interest of older juvenile and subadult females as a class in young infants (for example, DeVore, 1963; Jay, 1965). These reports have been supported by observations of the

Gombe troops and the Kampala caged group. In the latter group the young, nulliparous female mentioned above interacted extensively with all infants, often presenting and lip smacking to them, grooming them, and carrying them back to their own mothers. In the Gombe troops, young prepuberal and just postpuberal females were major participants in maternal subgroupings and interacted frequently with all infants. However, specialized greetings, grooming, and supportive relations, and the form and control of spacing patterns between particular members of such subgroups supported the assumptions made concerning familial and sibling relationships among these animals.

Male and female infants show consistent differences in both the development of the mother-infant relationship and interactions with peers as early as the second or third month. By the time of the transition period, sexual differences in play activity, in the frequency of initiation and withdrawal, and the duration and roughness of bouts, are already quite apparent. Such differences increase with age, and by the time of the birth of the next infant the young males have joined relatively permanent peer play-groups in which they spend a major portion of their time. Young females, on the other hand, tend to avoid rough and prolonged peer-group interactions and spend most of their time with the group of new mothers. During the next four to five years the males continue to interact mostly with each other on the periphery of the group, and generally are avoided by mothers and other females. In this same period the female juveniles maintain close proximity to adult females and attendant adult males, reach puberty, and perhaps experience their first pregnancies. Thus, at least until it is able to leave the mother-infant group, the young infant is surrounded by and interacts mostly with females. Such segregation may account for variations found in the nature of sibling relationships. For example, in the Gombe troops male juveniles interacted less than females with their known or presumed siblings, although indications of a strong bond between mother and juvenile were still apparent. During the first three to four months, then, a male infant may rarely have the opportunity to interact with older male juveniles, while a female has extensive experience with older females and the relationships among them. Thus the course and timing of gender-role learning through experiences with older individuals of both sexes will differ markedly for male and female infants.

The segregation of a baboon group by age and sex is incomplete because of the association of adult males with the maternal group and because of the occasional intrusion of male juveniles. One-to-two-year-old males, for instance, more often enter the maternal subgroup than older ones, and may participate in prolonged play sessions with the young infants. Older juvenile males occasionally join these sessions, but are much more likely to be chased off by an adult because an infant protests at their rough play, or to turn to interactions among themselves. In addition, three-to-five-year-old male juveniles devoted frequent attention to the female infants of young, low-ranking, and usually primiparous females, often carrying them off for grooming and play (Figures 7 and 8). The infants appeared to welcome the attentions of these juveniles, and would turn to them for protection and comfort in times of minor stress or in the absence of mothers and adult males. At the same time the male juveniles were experiencing their first extensive sexual and rudimentary consort behavior with their female peers. The development of these two interaction patterns may be closely related, as they appear to be in young hamadryas baboons (Kummer, 1968).

A final aspect of relationships among immatures to be considered concerns the occurrence of adoption. Observations in both cage and field indicate that the mother-infant bond need not have its basis in consanguinity to have extensive effects on the behavior and development of its participants. An orphaned infant baboon, hand-reared by one of the authors (Rowell, 1965), was introduced to the colony in Uganda at about three months, where it was quickly adopted by an adolescent female. Within a few days the relationship between this female and the infant was practically indistinguishable from that of a true mother-infant pair, and appeared to be developing in a normal manner. Thus this female's first infant effectively had a sibling. In the Gombe, a yearling male infant, orphaned by the accidental death of his mother, attached himself to the young, low-ranking female, Faith, prior to Charity's birth. Although the male showed no inclination to nurse, he spent much of his time in close proximity to Faith, was groomed frequently by her, and joined her in the maternal subgroup upon the birth of Charity. The adoption of a yearling female by a high-ranking adult male in Nairobi Park has been reported by DeVore (1963). Presumably such adoptions occur with some regularity among baboon troops, especially in areas where adult mortality is high, and add another dimension of variation and complexity to the structure and development of the immature's social environment and experience.

RELATIONSHIPS WITH OTHER ADULTS

Soon after the birth of her infant a new mother resumes an active role in the troop, and if possible spends most of her time in proximity to other mothers. Each Gombe troop, for example, usually contained at least one maternal subgroup, consisting of females with infants of about the same age, an entourage of small-to-medium-sized juveniles (some of whom were siblings to the infants), and one or more adult males. The composition of

Figure 7

Figure 8

Figure 9

Figure 10

such groups remained basically stable over considerable lengths of time, but varied from day to day according to circumstances within the troop. In addition, certain other classes of animals, such as cycling and pregnant females, older juveniles, and other adult males, occasionally joined these groups for brief periods of time. Thus the new infant is surrounded by a relatively permanent group of adults and immatures. On the appearance of a mother with a young infant an adult baboon may slowly approach the mother, showing various indications of friendly intent such as lip smacking, narrowing of the eyes, and flattening of the ears, and embrace her in a variety of ways depending on the mother's posture and the sex of the other adult. During this embrace the adult will touch the infant with hand or nose several times, and eventually may attempt to pull it out of the mother's arms. Such attempts for the most part appear to be intended to facilitate further investigation of the infant, but others are obviously motivated by a desire to carry it off. Usually the mother is able to retain her grasp, in which case the adult may only be successful in pulling the infant's rump out of her lap, which it may then sniff and lick before letting go (Figure 9).

Investigation of the ano-genital region apparently derives from the maternal pattern of care and cleaning of the infant. It continues to be a major form of interaction between the infant and other adults once the former has begun to leave its mother for brief periods. For example, an adult female may grab a passing infant, turn it upside down and inspect its genitalia and anus (Figure 10). That ano-genital inspection and manipulation itself is pleasurable is indicated by the fact that rather than struggle when handled in this manner, the infant usually ceases all activity and, if a male, may get an erection. The frequent inclusion of ano-genital orientation in adult social interaction patterns (presenting, mounting, handling of the genitals, and the like), usually linked with sexual behavior patterns, may derive from this investigation pattern.

The foot-back present is illustrative of a behavior ontogenetically derived from a specialized greeting behavior. Beginning as early as the fifth or sixth month an infant may start to initiate the lift-and-inspection sequence itself by approaching an older animal, turning around and walking backward up the other's chest such that its rear arrives at the level of the other's nose (Figure 11). This behavior pattern gradually becomes abbreviated and ritualized until at the age of twelve to fifteen months the immature may simply put one foot back to the groin or chest as it presents (Figure 12). Touching the foot in this manner is a common component of adult interactions, occurring in 50 to 60 percent of presents by females in the Gombe. It is also given by immature and mature males, but to a far less extent. This sexual difference may be attributable in part to the generally lower frequency of presenting by males, and to the development of an alternative pattern of interaction involving only them, the handling of the penis and scrotum, which is also first seen in adult-inspection of infants. This series illustrates the manner in which basic interaction patterns may be developed in the infant through consistent and positive experiences with adults.

The mother's reaction to the attempted attentions of another adult will vary with the latter's age, sex, and rank. The approaches of most males, especially those five to six years of age or older, are avoided, while those of young females are often ignored or simply tolerated. Interaction with another mother-infant pair, however, is sought after by mothers and occurs most frequently. In the Gombe and at Ishasha, extensive embracing and mutual investigation of each other's infants by mothers are frequently observed. Probably due to mutual interest in the infants and the relative security of the surrounding maternal subgroup, even the lowest-ranking females seemed fairly relaxed during such encounters and rarely avoided them. Frequent interactions between females whose infants were born about the same time appeared to be the basis for the development of strong, positive relationships, or maternity bonds, between them. As their infants mature, such females spend increasing amounts of time together (perhaps forming the core of a maternal subgroup themselves), groom frequently, and also support each other in aggressive encounters with others. The infants, in turn, play together more frequently than with others, and may occasionally use each other's mothers as bases for contact and support. One female of the mother pair may go off to feed or groom, leaving her infant in a play group that has formed around the other. On such occasions both infants may then return to the remaining mother for contact-reassurance and in times of stress. In one case a six-month-old female was even observed to mouth the nipple of the other mother on several occasions. Observations of enduring associations of young juveniles and adult females other than their mothers suggest that positive and preferential interaction between an infant and the female(s) with whom its mother shares a maternity bond may continue well past weaning and perhaps even into adulthood. Moreover, the age of the infants, rather than rank of the mothers, appears to be a major factor in the formation of maternity bonds, and this fact may be important in the ultimate determination of the infant's rank and status as an adult (see below).

All interactions between infants and other adult females, however, are not as positive as those described above. In both the Gombe and Ishasha groups many females responded to the mother-infant pair in an apparently ambivalent manner. On numerous occasions, for example, females were observed to embrace the mother and simultaneously attempt to prevent the infant from maintaining or establishing contact with her (Figure 13).

At other times, while grooming the mother, the female may poke, slap, and hit the infant in what appears to be a type of low-keyed, malicious play. The majority of interactions between infants and adult females without young babies of their own recorded at Ishasha were of this type. That such behaviors may represent a rudimentary form of rivalry between the female and the infant is suggested by additional observations: First, as suggested by Anthoney (1968), the embrace itself seems to be derived from the infantile ventral riding posture. Second, on many occasions yearling juveniles were observed to poke and slap at their young siblings when the latter were on the nipple (cf., observation notes on Myrna, Moley, and Loy, p. 128). Apparently, then, the mother's continuous attentions to her infant may produce emotions akin to jealousy in the other baboon with whom she interacts. Her reactions to such rough treatment of the infant will again vary with the individuals involved. She may try to limit or avoid such behavior from high-ranking females or from those with whom she shares a special bond, but rarely chastises them. Other females and immatures may eventually be attacked.

An infant may also experience aggression from other females at the time that it is being rejected and/or weaned by its mother. In the Gombe, females with whom the mother shared a maternity bond were observed on several occasions to threaten and push the infant away from its mother as it persisted in its efforts to attain the nipple. Mothers usually ignored such assistance, but sometimes they attacked the female briefly. On the other hand, adults in all our study groups often reacted to the end of a weaning bout, when the infant was finally accepted on the nipple, with a chorus of contact grunts and lip smacking, apparently indicating approval.

The importance of the mother's social status in determining the development of the infant's future role and rank within the troop has been partially documented in rhesus macaques by Koford (1963) and Sade (1968). Some observations on role formation in forest-living baboons may be added. Contrary to DeVore's (1963) report for savanna baboons, the rank of a female in our groups did not seem to be seriously altered by the birth of her infant. According to DeVore (1963, p. 316), "the protection afforded the mother (by adult males) while her infant is young is unique in baboon adult life and permits her to concentrate on the infant." In the Gombe troops, however, the nature of a mother's relationships with the adult males depended heavily on her social rank within the group. Mature adult males often sought close contact and extensive interactions with high-ranking females and their offspring, especially during the first six months. These males joined maternal subgroups containing such mothers for brief periods, and in some cases even became relatively permanent members of them. Adult males rarely showed comparable interest, however, in young

Figure 11

Figure 12

Figure 13

Figure 14

Figure 15

and low-ranking mothers. Such females only enjoyed the relative security and protection of proximity to an adult male by dint of their own efforts. When these efforts failed, such as when most or all of the males moved to the periphery of the troop during an agonistic interaction over an estrous female, these mothers were treated by the remaining animals according to their rank, despite the presence of their infants.

The contrasting examples of two previously mentioned females, Faith and Myrna, and their infants, Charity and Moley, will illustrate the important effect of rank on the experience of a mother and her infant. Prior to and during her pregnancy Faith appeared to be the lowest-ranking mature female in the troop. She was attacked fairly often, and without hesitation, by all other adult and even some adolescent females, and rarely received assistance from anyone. She was also occasionally attacked, and twice severely wounded, by the top-ranking adult male, Moon, during times of tension such as encounters between troops. Following Charity's birth, Faith attempted to stay close to another high-ranking male, Crease. Crease, however, was one of the more independent males of the troop and spent much of his time on its periphery. He interacted infrequently with Faith and Charity during the early months, and as a result they rarely enjoyed the relative protection of his presence. As early as the third week Faith began to be attacked again by large juveniles and other adult females as she attempted to avoid their attentions to Charity. During the twelfth week Crease began to show a special interest in Charity, and as a result the attacks on Faith became less frequent. By the twentieth week, however, Crease's interest had waned and he was once more rarely interacting with Faith and Charity. At the same time Charity was spending increasingly longer periods off her mother due to the latter's rejections. Then in the twenty-first week Charity herself was attacked for the first time by several juveniles and a young adult female during a tiff that developed out of a play session (Figure 14). Neither her mother nor nearby adult males came to her assistance on this occasion nor during two more attacks that took place in the remaining three weeks of the study. Thus it appears that Faith's inability to repel attacks successfully, either through her own efforts or by the support of an adult male, as well as her lack of support for Charity, have done much to determine the expectancies of others in the troop as to the infant's eventual status as an independent. Without further complication, then, it seems likely that Charity will be accorded a low rank like her mother.

The offspring of the high-ranking female, Myrna, had quite different experiences. In accord with her rank, Myrna enjoyed extensive supportive relationships with other females; and frequent and close contact with her, Loy, and Moley was sought after by a number of males in the troop. On the rare occasions that she was attacked by other females or adult males, someone invariably came quickly to her assistance, usually the adult male Harry with whom she shared a close and intense relationship. As mentioned above, Moley enjoyed the protection and support of his sibling, Loy, as well as that of Harry and Myrna. This consistent and obvious support was acknowledged by older juveniles and other adult females by the hesitation with which they approached and interacted with Moley, and by the frequency with which they presented to him and/or Myrna prior to such interactions. The effect on Moley of this support and the deference it produced could be judged by the confidence and relative immunity with which his sister Loy behaved in her interactions with peers, older females, and even adult males.

Differences in early experience such as those seen between Moley and Charity may tell us a great deal about the development of status and rank in a group of primates. It is possible that the effects of such early experiences will be counterbalanced by additional relationships with others in the troop. For example, Charity's associations with several young male juveniles, and with the adult male Crease, which will be described below, may bring about an enhancement of her rank when she reaches puberty.

Some outstanding and persistent associations between infants and adult males in the forest-living troops of the Gombe National Park have been reported elsewhere (Ransom and Ransom, in press) and will be mentioned briefly here. These male-infant associations have particular significance for the process of social development because of their potential contribution to the bonds that were observed among the adults in the Gombe troops. The following is a brief summary of four types of relationships that were observed to develop between males and infants.

FOUR TYPES OF MALE-INFANT RELATIONSHIPS

The first type of relationship consisted of an intense and long-lasting association between an adult male and the offspring of a particular female. In the Gombe this relationship derived from the pair bond shared by the male and the infant's mother, such as that between Harry and Myrna. Harry fulfilled a paternal role with respect to Loy and Moley that included "baby-sitting" for them (that is, remaining with an infant off its mother, grooming and sometimes carrying it, and providing frequent, reassuring contact), and providing passive and active protection for them throughout the duration of the study. His association with each infant appeared to be strongest during the early months, but lasted well beyond the weaning period. After Moley's birth, Loy continued to seek and receive contact and protection from Harry during times of stress, an interaction pattern he shared with no other juvenile to such a degree.

3/25/69 10:52 A.M. Myrna, Moley, and Loy are feeding in a twenty foot palm tree at the base of the path. Harry is sitting on the path about forty feet to the east, I am just off the path to the north of the tree, there are no other baboons in sight. Myrna with Moley ventral suddenly shoots down out of the palm, dropping the last six feet to the ground, Moley squeaks as she lands. As Myrna runs off southeast, Loy leaps from the palm to another to the west, but misses and falls to the ground about fifteen feet from me. As she is falling, Harry charges straight toward me, threatening and pant-grunting. Loy lands hard, obviously gets the breath knocked out of her, and moves slowly off to the east, squeaking softly. As Harry approaches, I turn and walk away, he follows still pant-grunting and with full piloerection. He eventually stops and sits down as Loy disappears to the east and I move farther west down the path.

The extent of the association between male and infant depended on the nature of the relationship between male and mother which, in turn, varied with the latter's age and rank. Most young and low-ranking females appeared to lack a well-developed pair bond, and consequently their infants did not benefit from an association with a male to the degree that Loy and Moley did. Variations of this sort in the early experience of first- and later-born infants, and of infants of low- and high-ranking mothers, may well combine with others mentioned above in determining the nature of the role and rank that the infant assumes, and is accorded, as an adult.

A second kind of relationship between adult males and infants is derived from the male's general role of protector within the troop. During moments of sudden or extreme tension such as at the appearance of a predator, all animals, but especially adult males, may grab up and carry off an unattended infant. In the Gombe the nearest male at hand in such situations was most likely to respond in this manner; for some infants this was most often the male with whom the mother shared or was developing a pair bond. However, in some cases such interaction occurred repetitively between an infant and a male who shared a no more intense relationship with the mother than co-membership in the same major subgroup. The most notable example involved a mother who apparently showed little concern for the welfare of her young daughter. As early as the first week this mother often left the infant alone or with immatures while she went off to feed and groom, and was consistently slow in responding to the infant's signs of distress. A high-ranking male responded to this treatment, and to the proximity of interested chimpanzees, by staying with and often carrying the infant on his chest for extended periods. The chimpanzees actually made several attempts to capture the infant, but were thwarted in this case by the male's attentions. Further development of the association was interrupted by the death of the infant from unknown causes at the age of three months. Neither before the infant's birth nor after its death did the male show any special attention to the mother herself. Thus such an association may be the basis for a bond whose formation is independent of any existing ones in the group.

A third type of male-infant association observed in free-ranging troops of baboons also seemed to be independent of existing relationships between adults. It involved the frequent interaction of young males (approximately four to eight years old) with the infants of young, low-ranking females, such as Charity. These interactions, which included patterns of play, grooming, and sexual behavior, were highly reminiscent of those that appear to be involved in tripartite relations and the formation of one-male units among hamadryas baboons (Kummer, 1967). In the Gombe troops, young males began to attend to and occasionally carry off these young infants as early as the second month, and eventually the infants began to initiate the interactions themselves. The young mother's inexperience and lack of well-established pair and subgroup relationships, and the consequent availability of the infant, made the development of these associations possible. Such associations, especially those involving the younger males, may well be the basis for developing pair bonds and consort preferences.

The fourth and most outstanding relationship observed was a feature of dominance interactions among adult males, and consisted of the tendency of some males to grab up and carry infants under conditions of stress. The adult male seemed to enhance his effectiveness in interactions with other males, insofar as close contact with an infant seemed to inhibit aggressive behavior from them. Similar behavior apparently designed to produce special status has been observed among subordinate or immature males both in hamadryas baboons (Kummer, 1967) and in Japanese macaques (*M. fuscata*) (Itani, 1963). In the Gombe and Ishasha troops, however, high-ranking males were observed to use this technique as well as immatures. The most frequent stimulus for the behavior in the Gombe was the presence of young, mature males who were in the process of transferring from one troop to another. The particular association of male and infant seemed to depend on the nature of the male's rank and status in the group and the availability of infants at the time. One such association developed between the top-ranking male, Moon, and Myrna's son, Moley. Moley was born about two months after a large young male, Grinner, had begun his transfer to the troop. Moon paid great attention to Moley from the day of birth, often taking him from Myrna, and carrying and grooming him. These thefts seemed designed in part to ensure the proximity of the high-ranking mother, as Moon would often only allow Myrna to retrieve her son if she consented to groom. During the second month Moon began to time these interactions with the infant to coincide with his highly tense status interactions with Grinner. Another high-ranking male, Crease, reacted to Grinner's presence in a similar fashion, but apparently because of the lack of young infants in his subgroup, chose a yearling male, Icarus. Somewhat later when new infants, with their black natal coats, appeared in his group, Crease began to use one of them, Charity, in preference to Icarus.

The form and effect of such an association with a male varied considerably with the age of the infant involved. The enforced separations from his mother and the rough treatment received at Moon's hands were obviously quite stressful to Moley (Figure 15), and as he matured he tended to avoid contact with Moon and to seek Harry's company. By the time of his association with Crease, however, Icarus was spending most of his time away from his mother in peer-group interactions and rough play with older immatures. He seemed to welcome Crease's attentions, and eventually began to initiate the sequence himself by approaching either Crease or Grinner and indicating a high degree of excitement. Moreover, as Crease began to associate with Charity, Icarus started to approach other adult and subadult males in the same fashion, and even tried to initiate the sequence with reference to Grinner when backed up only by his peers.

That this type of association with an adult may serve at least as a prototype for the infant's future relationships, including consort preferences and supportive bonds, is suggested by several lines of evidence. Rudimentary forms of the male-infant-male interaction pattern are seen among juveniles and infants in immature mock battles. Subadult males were also seen to form such associations with older yearlings and young juveniles in apparent, usually unsuccessful attempts to intimidate older males. Finally, the behavior patterns seen in adult male supportive interactions, which include close ventro-ventral or dorso-ventral embraces, are highly reminiscent of those seen in the male-infant-male pattern.

Although individuals were not often recognized at Ishasha, interactions recorded between infants and adult males all fitted descriptions of types 1 (pair-bond male), 3 (male protector intensified), or 4 (rank-enhancing male), male-infant interactions seen at Gombe stream. In most cases, there were presumably individual relationships that were not recognized by the observer, but the possibility remains that these types of interactions may be definable without reference to individual relationships.

CONCLUSION

In the preceding pages we have described part of the variety and range of relationships and interactions that the young forest-dwelling baboon may experience. As the examples demonstrated, the actual experience of any one infant is an outcome of the interaction of a complex set of factors peculiar both to the mother-infant pair and to the circumstances within the social group as a whole. Some of these factors appear to be basically predetermined at birth, while others are an outcome of the temporary combination of events and relationships within the troop. The physical and social environment of the infant is at first provided by the mother. Her age, rank, and previous maternal experience determine to a great extent the primary format for the mother-infant interaction pattern, as well as the nature of the infant's first experiences with others, especially siblings and adult males. At the same time, the composition of the social group and the particular balance of relationships among its members determine the nature of the social environment within which the infant matures.

There is enough variation in the early experience of infant baboons to provide most of the explanation for the comparable variety seen among adults and their relationships. However, at this point our understanding is far from complete, both of the extent of possible variation, and of the relative contributions of the multiplicity of factors to the infant's social development. For example, in the Gombe the offspring of two females, Faith and Myrna, experienced very different relationships with their mothers and others in their infancies. These experiences, and the consequent relationships that these young baboons developed in the group, appear to have laid the groundwork for the development of adults who will enjoy radically differing social rank and status. However, as they pass through adolescence and early maturity they will be subject to continuing influences from the group, and the effects of previous experiences and relationships might be partially or even completely offset. Charity's relationships with the adult male Crease and with a number of young males in the troop could be the basis for important adult bonds, such as consort relationships and pair bonds, and thus may provide her with the means for escaping from the legacy of her low-ranking mother. The task of discovering the factors and the intricate combinations that direct the form and development of social behavior has been begun in a number of field and cage studies of primates, including those of baboons reported here, but so far one of their main results has been to emphasize that a great deal more long-term observation, experimental manipulation, and analysis of primate behavior and group social structure in both field and cage situations will be well worthwhile.

References

ANTHONEY, T. R. (1968) "The Ontogeny of Greeting, Grooming and Sexual Motor Patterns in Captive Baboons (Superspecies: *Papio cynocephalus*)." *Behavior,* 31: 358–372.

BRAMBLETT, C. (1967) "Pathology in the Darajani Baboon." *Am J. Phys. Anthrop.,* 26: 331–340.

DEVORE, I. (1963) "Mother-Infant Relations in Free-ranging Baboons." In H. L. Rheingold (ed.), *Maternal Behavior in Mammals.* New York: Wiley, 305–335.

HALL, K. R. L. (1963) "Variations in the Ecology of the Chacma Baboon, *Papio ursinus.*" *Symp. Zool. Soc. Lond.,* 10: 1–28.

——— and I. DEVORE (1965) "Baboon Social Behavior." In I. DeVore (ed.), *Primate Behavior.* New York: Holt, Rinehart and Winston, 425–473.

HINDE, R. A., T. E. ROWELL, and Y. SPENCER-BOOTH (1964) "Behavior of Socially Living Rhesus Monkeys in Their First Six Months." *Proc. Zool. Soc. Lond.,* 143: 609–649.

——— and Y. SPENCER-BOOTH (1967) "The Behavior of Socially Living Rhesus Monkeys in Their First Two-and-a-half Years." *Anim. Behav.,* 15: 169–196.

ITANI, J. (1963) "Paternal Care in the Wild Japanese Monkey, *Macaca fuscata.*" In C. H. Southwick (ed.), *Primate Social Behavior.* New York: Van Nostrand, 91–97.

JAY, P. (1965) "The Common Langur of North India." In I. DeVore (ed.), *Primate Behavior: Field Studies of Monkeys and Apes.* New York: Holt, Rinehart and Winston, 197–249.

KAUFMANN, J. H. (1966) "Behavior of Infant Rhesus Monkeys and Their Mothers in a Free-ranging Band." *Zoologica,* 51: 17–29.

KOFORD, C. B. (1963) "Rank of Mothers and Sons in Bands of Rhesus Monkeys." *Science,* 141: 356–357.

KUMMER, H. (1967) "Tripartite Relations in Hamadryas Baboons." In S. A. Altmann (ed.), *Social Communication Among Primates.* Chicago: University of Chicago Press, 63–71.

———. (1968) *Social Organization of Hamadryas Baboons: A Field Study.* Chicago: University of Chicago Press.

RANSOM, T. W. (1971) "Ecology and Social Behavior of Baboons in the Gombe Stream National Park." Ph.D. dissertation, University of California, Berkeley.

——— and B. RANSOM. "Adult Male-Infant Relations Among Baboons (*Papio anubis*)." *Folia Primat.,* in press.

ROWELL, T. E. (1965) "Some Observations on a Hand-reared Baboon." In B. M. Foss (ed.), *Determinants of Infant Behavior,* Vol. 3. New York: Wiley, 77–260.

———. (1966) "Forest Living Baboons in Uganda." *J. Zool. Lond.,* 149: 344–364.

———. (1967) "A Quantitative Comparison of the Behavior of a Wild and a Caged Baboon Group." *Anim. Behav.,* 15: 499–509.

———. (1968) "The Effect of Temporary Separation from Their Group on

the Mother-Infant Relationship of Baboons." *Folia Primat.*, 9: 114–122.
———. (1969) "Long-term Changes in a Population of Uganda Baboons." *Folia Primat.*, 11: 241–254.
———, N. A. DIN, and A. OMAR (1968) "The Social Development of Baboons in Their First Three Months." *J. Zool. Lond.*, 155: 461–483.
SADE, D. S. (1965) "Some Aspects of Parent-Offspring and Sibling Relations in a Group of Rhesus Monkeys, with a Discussion of Grooming." *Am. J. Phys. Anthrop.*, 23: 1–17.
VAN LAWICK-GOODALL, J. (1968) "The Behavior of Free-living Chimpanzees in the Gombe Stream Reserve." *Anim. Behav. Monog.*, 1: 161–311.

Yukimaru Sugiyama

Social Characteristics and Socialization of Wild Chimpanzees

INTRODUCTION

PHYLOGENETIC AND ECOLOGICAL SETUP

The habitat of chimpanzees (*Pan troglodytes*) includes the tropical African rain forest and the surrounding savanna woodland. Their distribution ranges from longitude 15° West to 32° East and from latitude 120° North to 8° South. The pygmy chimpanzee (*P. paniscus*) found to the south of the Congo River is somewhat different from other chimpanzees, and is separated from them by the Congo River. Three subspecies of chimpanzees, *P. t. troglodytes, P. t. verus,* and *P. t. schweinfurthii,* are separated from each other by the Niger and the Ubangi rivers, but the morphological variation among these three subspecies is slight.

Chimpanzees were long considered to be thick-forest dwellers, but recent studies on their distribution and ecology reveal that population density is greater at the forest edge than in the interior of the high forests, and that they prefer the complex habitat of forest, woodland, and savanna (Suzuki, 1969; Kano, in press). Some behavioral and sociological studies indicate that the chimpanzee originated in the savanna woodland, adding the high forest to its habitat later on (Kortlandt, 1963; Itani, 1967; Itani and Suzuki, 1967).

Chimpanzees mainly eat fruits and nuts. Feeding habits and movement patterns vary seasonally, depending on the change of the fruiting season of each plant species in their habitat. Although chimpanzees are generally frugivorous, forest dwellers are rather baccivorous, while savanna-woodland dwellers are rather graminivorous (Reynolds and Reynolds, 1965; Sugiyama, 1968; Suzuki, 1969). Chimpanzees also eat termites, some other kinds of insects, and occasionally meat, such as monkeys and young terrestrial mammals (Goodall, 1965; Kawabe, 1966; Suzuki, 1969b).

Chimpanzees are phylogenetically closest to man; they inhabit the wide area of African forest and savanna woodland where man's ancestors probably originated, and they possess a highly developed behavioral pattern that suggests possible analogies to ancestral hominid behavior. Thus, the behavior, sociology, ecology, and life history of wild chimpanzees are among the most important themes adopted when we search for and reconstruct the origin of man and the details of the behavior, sociology, ecology, and life history of ancestral man.

The main data on chimpanzee ecology and sociology treated in this article are from the author's 360 hours of observation between September 1966 and March 1967 of the wild chimpanzees that primarily inhabit the edges of the Budongo Forest, Uganda, East Africa. Details of the ecology and sociology of these chimpanzees are found in his original articles (Sugiyama, 1968, 1969).

OUTLINE OF SOCIAL LIFE

For most primates the troop, with its constant membership, is the most stable social unit. Each troop has its own home range of which the main part, the core area, is separated from others, even though home ranges sometimes overlap. The home range itself, or the core area of some species, is exclusively occupied by a troop as a territory (Sugiyama, 1968).

Chimpanzees, however, lack a stable social troop. Our many observations have revealed that chimpanzees temporarily gather in a party to move together. Such a unit, however, frequently splits into smaller parties or solitary animals. From the viewpoint of social composition, most of the parties can be classified as "mother parties," consisting of mothers and babies; "male parties," consisting of active adult and subadult males; "adult parties," consisting of adult and subadult males and females; and "mixed parties," containing animals of many ages and sexes. But no stable social bond among individuals, except the mother-infant dyad, characterized the chimpanzees of the Gombe Stream Reserve (van Lawick-Goodall, 1967).

Recently with the method of individual identification, extended observations revealed that chimpanzees who move in a certain area are comparatively constant, and although they don't form a strictly organized troop, they are loosely connected to each other by their common moving range. Chimpanzees form loose flexible social units, termed regional populations, which are loosely separated from other neighboring regional populations (Sugiyama, 1968).

In the author's main research area of the Budongo Forest, a regional population of fifty to sixty chimpanzees, Regional Population A (RP-A), occupied about 7.5 square kilometers. (According to A. Suzuki, who succeeded the author at his research area of the Budongo Forest, the regional population of this area consisted of about eighty chimpanzees covering nearly 20 square kilometers.) Occasionally most of the chimpanzees of this population formed a party that move together. Usually, however, chimpanzees formed parties of less than ten or moved individually. They also gathered into one party that subsequently divided into two or more. The population density of the Budongo Forest was about six to seven animals per square kilometer. The moving range of RP-A was surrounded by moving ranges of neighboring regional populations. These ranges frequently overlapped. Sometimes members of RP-A formed a mixed party with members of RP-D where their home range overlapped, but the mixture of members with those of RP-B or RP-C was never observed. Probably the differences in the social relationships between regional populations depend on the geographical and vegetational isolation or historical factors; for example, RP-A and RP-D might have divided into separate populations from one.

DEVELOPMENT OF YOUNG CHIMPANZEES IN THEIR SOCIETY

THE INFANT 1 AND INFANT 2 STAGES

For the first several months (infant 1 stage) a newborn almost always nurses and sleeps clinging to its mother's ventrum. As it matures, after ten weeks from birth, it begins to stretch its hand, touch a nearby twig or leaf, pull near to it and then occasionally put the twig or the leaf into its mouth. All the while the youngster grasps its mother with its hands or feet while she sits and feeds or rests calmly. After a few months the infant leaves its mother to explore the ground or a nearby branch. Its first attempts to leave its mother's bosom are always curtailed by the mother, but gradually she allows the infant to leave her control.

A yearling (infant 2 stage) always moves with its mother, usually clinging to her ventrum or riding on her back. When the mother sits at a resting or feeding site, the yearling leaves to play alone within a few meters of her. Generally speaking, young chimpanzees are very curious and approach the observer to watch and observe him. The infant's subsequent behavior depends on the mother.

Mama (adult female) became accustomed to the author early during the study, and when her child, Mtoto (infant 2), approached the author at the top of the branch several meters away from its mother, she continued to eat calmly. So Mtoto also quickly became accustomed to the author. But Mgeni, Ndizi, Ngabi, and other mothers ran away from the observer or hid behind branches and leaves when they found the author under the tree. Their infants reacted likewise. The observation of those infants was rather difficult throughout the study period. The habituation of infants to the author mimicked that of their mothers, as they directly copied their mother's attitude toward him. In this way youngsters may learn that the leopard is dangerous, and, on the other hand, that the elephant is not.

Chimpanzee youngsters also mimicked their mothers' attitudes toward other chimpanzees. For example, a chimpanzee youngster will approach another chimpanzee to whom its mother manifests an associative attitude. But it will avoid a big male to whom the mother exhibits submissive behavior. However, a chimpanzee primarily exhibits associative behavior to any chimpanzees of its regional population, whereas a macaque, for example, conspicuously changes its attitude toward troop members according to the status relationship between them.

Infants apparently learn the food repertoire by watching what their mothers eat; infants commonly sample the foods that their mothers consume. Infants are allowed to bite or grab at food in their mother's mouth or hand. A mother will also reward the begging gestures of her infant by sharing food with it. Adult chimpanzees consume four or five ripened figs (*Ficus mucuso*), chew the juice, and make fig knead. Because of the size of the fig, however, infant chimpanzees cannot copy this behavior. One day Mtoto was looking at its mother who was eagerly chewing figs and making knead. The infant begged for some by stretching its hand. Mama took the fig knead from her mouth, divided it into halves, and gave one-half to her child. Mtoto took it in its hand and put it into his mouth to eat. By repeating such behavior or similar exercises, and by copying the mother's behavior, infants may learn the behavioral patterns necessary for survival.

Chimpanzee mothers seldom consciously gather together, and usually a mother is observed moving alone accompanied only by her offspring. Occasionally, however, mother groups gather at a fruit-bearing tree, because of their common preference for a certain kind of food and/or because their common low activity leaves them behind the quick and active movement of males and single females.

When mothers gather at a fruit-bearing tree, infants come near each other. Social interactions among infants, such as simple group behavior, casual chasing, and nonvigorous wrestling were observed when a feeding or resting party gathered at a tree. However, they were infrequent at the research area of the Budongo Forest, since same age infants did not frequently gather in a party. Usually an infant was found playing alone near its mother or being cared for by an elder juvenile or a subadult.

From the observations of Japanese macaques (*Macaca fuscata*), the traditional food habits are rigidly maintained by elder members of the troop, but new habits, for example, the potato-washing habit, are acquired by young members. Such newly acquired habits are propagated at first from a discoverer-inventor to its playmates; secondly to their mothers, brothers, and sisters; and then, gradually, to other members of the troop (Itani, 1958; Kawai, 1965).[1] While a mother plays an important role in a youngster's learning the traditional group habits, young playmates play an important role in a youngster's acquiring new habits during development. In chimpanzees as well as in Japanese macaques learning from playmates must be as important as learning from the mother, not only for the youngster's survival but also for learning daily routines.

As the infant matures, sessions of social grooming by its mother become increasingly longer and probably serve to reinforce their social bond when the infant returns to its mother after leaving her for a longer period of time (van Lawick-Goodall, 1967). During infancy chimpanzees remain close to their mother, strengthening the social ties with her, and copying from her those behavioral patterns that are necessary for survival. This process can be called the first stage of socialization.

MOVEMENT OF JUVENILES

Even though juveniles (animals between two and six years old) are far more active and frequently play in groups as opposed to infants who don't, there is little opportunity to gather with other juveniles at a tree as their mothers don't consciously congregate in a party. In the chimpanzees' flexible society, where each individual frequently joins a party that subsequently divides into separate parties, social interaction among age-mates is infrequent. On the other hand, in the many monkey species that are characterized by a stable and rigid troop organization, especially those of macaques living in large bisexual troops containing many age-mates, many infants and juveniles frequently gather in play groups.

Social interactions among macaque juveniles are rigidly restricted by the mother's social status. A juvenile Japanese macaque born to a peripheral mother in a large troop plays with juveniles of other peripheral mothers but rarely interacts with infants of central dominant mothers (Koyama, 1965). On the other hand, in the flexible and associative chimpanzee society a juvenile interacts socially with any juvenile of its population, although such interaction may be infrequent. Sometimes a juvenile chimpanzee even interacts peacefully with a juvenile of the neighboring population. Such behavioral characteristics must be related to the following social characteristics. Japanese macaques are tightly connected with each other in a troop the members of which are antagonistic to members of the neighboring troops, and with solitary monkeys. Even

[1] Editor's note: Poirier (1969) discusses a similar system among Nilgiri langurs.

in a troop, according to the spatial distribution of its individuals, an individual is tightly connected with only blood-related individuals and with others of a similar status, but has little social interaction with other troop members.

During the juvenile stage a chimpanzee first attempts to build its own nest (Goodall, 1962), and at the late juvenile stage it occasionally moves independently of its mother. In reacting to the booming noise of chimpanzees of another party of the same regional population, resting or feeding chimpanzees frequently run toward the sound, also making loud booming noises. Mothers with babies and other individuals wishing to continue feeding or resting remain, and so, a party divides into two. In such circumstances, sometimes a juvenile follows the active moving party of elder chimpanzees and joins another booming party, leaving its mother behind.

Juveniles gradually learn to move independently of their mothers and move with their age-mates and familiar elders. The juveniles not only play and travel in such groups, but it is also in these groups that they learn to search for fruit-bearing trees and to avoid predators. While separated from the mother, young chimpanzees travel with other individuals of the same regional population, that is, individuals acquainted with each other and among whom there is a feeling of well-being and assurance.

Chimpanzees usually move in parties of less than ten, or alone. While most solitary-moving chimpanzees are full-grown adults, several times senior juveniles between four and six years old, both male and female, were observed moving alone in the forest. They move calmly without fear, and even after encountering the author they didn't flee. However, senior juveniles infrequently traveled alone.

Most of the Japanese macaques, especially the females and the young, depend upon the protection of the troop. If a monkey loses its troop, it becomes very uneasy. Individual Japanese macaques usually move with the troop, even if they wish to move in an opposite direction from it. Some adult males desert the troop and spend months and even years by themselves and are subsequently refused reentry into the troop (Nishida, 1966). On the other hand, the social environment of chimpanzees allows a youngster to move freely with little restriction; thus each individual may be able to develop and establish its own individuality.

CHARACTERISTICS OF GROUPING ACTIVITY

SOCIAL RELATIONS AMONG PARTY MEMBERS

A rigid dominance hierarchy is characteristic of most primates, especially of the adult males. Each animal knows its position in relation to others and behaves accordingly. The dominance hierarchy functions to reduce interanimal friction and antagonism; it reduces intratroop conflict to a minimum. The ranking relationship between individuals is frequently exhibited in daily social behavior. Encounters between two closely ranked adult male Japanese macaques are characterized by behavioral sequences that cannot be readily classified as dominant or subordinate actions. But if one animal is clearly dominant and the other clearly subordinate, the dominant animal ignores the subordinate one upon meeting, while the latter immediately manifests a clearly subordinate behavioral expression. If the distance between the two is minimal, the dominant acknowledges the latter and tries to manifest its rank through subtle means. The subordinate then exhibits both dominant and submissive gestures.

In chimpanzees dominant and subordinate relations are subtle, especially in relationships between males of similar size. This is shown in the following example. An adult male was feeding on a branch of a fig tree. He fed calmly as another male approached him in the same branch that was too narrow for the second male to pass the first male without touching him. Then, the second male stepped over the back of the first, who continued to feed. Both males didn't hesitate to approach each other, doing so peacefully. Because the figs (*Ficus mucuso* and *F. capensis*) grow in clusters concentrated at the base of big branches or the tree trunk, the best feeding places are those restricted to small areas. In such a tree, two full-grown adult males were sometimes observed feeding zealously with their bodies touching each other but without exhibiting dominance behavior. This kind of tolerance among individuals may be related to the fact that any individual can move independently from the others and can easily join any party of its regional population. There are few social pressures restricting individual behavior.

Sometimes mild dominance interactions are witnessed between a big adult male and another smaller male. For example, once when a small male fed while sitting on a narrow branch, a big male sat near him and began to eat. The small male stopped eating and left expressing only minor submissive behavior. Chimpanzees may coexist without a strictly ordered ranking hierarchy, even between a big and small male. Chimpanzees are remarkably tolerant of each other when compared to macaques who are characterized by a rigid dominance hierarchy.

Among many primates, when two individuals meet, the high tension

Figure 1. A full-grown male, Mzee, eating figs on a *Ficus capensis* tree.

between them is usually eased by either defensive or adulated behavior or expression. Grooming is most frequently the friendly or associative behavior used to ease such tension. Chimpanzees exhibit many kinds of friendly and associative behavior patterns (besides grooming) that are used for appeasement or to ease social tension. When excited party members produce the booming noise and the social tension among individuals is high, when two parties meet at a fruit-bearing tree and tension between both party members becomes high, when two individuals come too near on a narrow branch, or in other situations, one grooms the other, one stretches its hand and touches the other's body, one grips or faces the genital area of the other, one embraces the other, or some other behavior may occur. These greeting behaviors are not always forms of appeasement directed by a subordinate to a dominant individual. In many situations, when two or more individuals meet, tension-easing behaviors occur on an equal basis, independent of the actions of the dominant and subordinate relationship. These interindividual relations suggest that the dominance hierarchy functions subtly to regulate social friction among chimpanzees.

Observations of chimpanzee mating behavior confirm the view that chimpanzee society is not strictly dominance oriented. The sexually excited male and estrous female chimpanzee do not form an extended consort couple, nor do they copulate exclusively. An estrous female may mate with many males within a short time, even in sight of other party members. Thus sex may be potentially disruptive, that is, as it is in baboons and macaques. Some estrous females mated with several males in a party, including young adult males. An estrous female once stopped grooming an adult male, approached a young adult male on a nearby branch and copulated with him before returning to the big male. These examples show that sexual relationships, as well as other social relationships among chimpanzees, are formed without the constrictions of a rigid dominance hierarchy.

Although the rigid dominance hierarchy in the Japanese macaque society makes possible the coexistence of many individuals in a small area without overt fighting, it controls and suppresses an individual's desires in preference for maintaining an organized and united troop life. For example, a young male macaque must always be aware of a possible attack from a dominant male when he mates with a female. Thus he unobtrusively mates with a female for only a short time on the troop's periphery. On the other hand, in the chimpanzee society even the mating of a young or subordinate male is rarely disrupted by a dominant male. The difference between these two primate species must be dependent upon their social organization. Youngsters learn the nature of their particular social organization through species-specific socialization practices.

Although the author has emphasized the uniqueness of the chimpanzee society in the friendliness and the associativeness among individuals, inter-individual friction is not always solved by friendly and associative behavior. Sometimes a big male attacks and chases a small male, pulling the whole party into excited confusion. This is especially common during the dry season when chimpanzees form large parties that noisily travel over a wide area. But the dominant and subordinate relationship does not govern the whole of their daily life. The author observed only thirty-one dominant and aggressive-submissive behaviors among chimpanzees of RP-A during his six-month study. Each individual can move alone if it does not wish to follow the mandates of others.

Chimpanzees exhibit much individuality in their friendship preferences and in their moving patterns. Some animals seem, strictly on the basis of personal preference, to associate more with some animals than others. Thus they mimick human beings. Chimpanzees differ from each other in their physical, social, and psychological make-up, and this might have paved the way for the establishment of a loosely knit social organization. The loose structure of the chimpanzee social organization permits a wide range of freedom of choice for each animal.

FELLOWSHIP AMONG REGIONAL POPULATION MEMBERS

Excited chimpanzees frequently produce the booming noise, that is, they bark agitatedly, drum tree trunks and buttresses, run and brachiate from branch to branch, and vigorously shake branches. In this context the booming noise communicates with other chimpanzees and functions to draw them together. Chimpanzees within hearing range of the separated party, and who belong to the same regional population, frequently respond similarly and run in the direction of the sound. The two booming parties often meet and join at a fruit-bearing tree. Even though chimpanzees of a regional population cannot visually recognize each other in the thick forest, they can vocally communicate the existence of fruit-bearing trees, predators, or their own location and demeanor (Reynolds and Reynolds, 1965; Sugiyama, 1969).

The booming clamor of chimpanzees occurs when a party divides and half the party moves away, when two parties meet, when a party finds or reaches a fruit-bearing tree, and when a party begins to travel after feeding or resting. It also occurs prior to the active morning and evening feeding periods. When the booming noise of distant chimpanzee parties is heard, and when a party is traveling quickly, it also responds boisterously (Sugiyama, 1969). The booming clamor functions to communicate with regional population members and, as such, strengthens their social bond. Although the evidence is limited, the author was impressed by the fact that chimpanzees may raise a clamor just for the sake of expressing excitement. The excitement of one chimpanzee stimulates excitement in others, and the entire party soon becomes engulfed in frenzied excitement through a feedback process.

Many tribal peoples exhibit the trance and possession state as typified by the trance dance of the Kalahari bushmen (Lee, 1968). Although the Kalahari bushmen possess a conscious rationale for performing the trance dance—to remove an evil spirit from a sick person—they dance because of the psychological gratification derived from the intense excitement and lapse into unconsciousness. Chimpanzees certainly don't have a conscious rationale to explain their booming clamor, but they may share a common underlying psychological motive or drive with the trance and possession state of some tribal groups of man.

Having contact with each other via these communicative behaviors, regional population members gather and join at fruit-bearing trees; on the other hand, they split and divide at will. The regional population is a group of chimpanzees who are acquainted with each other and who frequently move together, although not always as a unified troop. For each individual the bond with its home range must be strong, and based on this bond, a young chimpanzee may learn to identify neighbors and strangers during its development. Gradually it may become conscious of a fellowship with chimpanzees of the same regional population.[2]

In the party of RP-A, the author sometimes found strangers who appeared in his research area only once or twice throughout the study period. Though they were not positively identified, these strangers might have wandered into the range of RP-A from the neighboring regional populations of RP-B, RP-C, or RP-D, and might be slightly acquainted with the research area and with the resident chimpanzee. When a lone stranger joined a party of RP-A members it did so calmly, and the residents neither showed antagonism nor attempted to exclude it. The social relationships between residents and a visitor differ from the social relations between troop members and a solitary male in many primate species. In many species, a solitary male is strictly forbidden by resident troop members to wander into their area. However, at the overlapping area of the home ranges of two regional populations, the author twice observed that many chimpanzees of both regional populations met, excitedly mixed together,

[2] Editor's note: Poirier (1968) discusses a similar "bonding" mechanism to the home range, and to fellow group members, among Nilgiri langurs.

exaggeratedly ate leaves and fruits, ran and leaped about the ground and branches, drummed and beat the buttresses of the huge trees, and uttered their desperate cry and bark. After about one-half hour of noisy booming clamor, chimpanzees of each population returned to their own home range. During this time no display against members of the other population was observed. Although the author could not interpret the exact meaning of these excited and exaggerated behaviors, social relations of neighboring populations may be peacefully maintained by such "ritual" behavior.

The author was impressed by the fact that each chimpanzee living in the continuous large forest recognized members of its own regional population and that nonantagonistic relationships with chimpanzees of neighboring regional populations existed. Not only do smooth, friendly social relationships stem from this, but genetic exchange among regional populations is also quite likely.

Chimpanzees at Kasoge, the savanna woodland to the northeast of the Mahali Mountains in Western Tanzania, were baited at the overlapping part of the home range of two regional populations by Nishida. These chimpanzees move in a temporarily formed party or by themselves as do those of the Budongo Forest. Frequently one party joins another party that may divide again into two. But scores of chimpanzees are loosely connected with each other and a party is formed from these members. Usually a regional population of about thirty chimpanzees, the Kajabara group, occupied the study spot. But when another regional population of about fifty chimpanzees, the Mimikire group, came to the study spot from its proper home range on the southern side during the dry season, the former as a whole avoided the latter and moved to the northernmost part of its home range, repeating the division and confluence of parties in their daily life. Between 1967 and 1970 only four females of the Kajabara group remained in the study spot and shifted to the Mimikire group when the Kajabara group deserted the place. These females followed the Mimikire group as far as the southernmost limit of its home range when it returned there during the wet season (Nishida, 1968; Kawanaka, in press).

Although the group pattern among the Budongo chimpanzees is described as loose, variations exist. For example, while chimpanzees of RP-C were sometimes observed to move as a party, the chimpanzee groups of RP-A and RP-D rarely did so. Although variations occur between the different regional populations of the Budongo Forest in the amount of independency that is uniform among the regional populations, the patterns are still all loose when compared to the group structure of macaques, for example, which is much more rigid. Kasoge chimpanzees display less variability and are more rigid than Budongo Forest chimpanzees. Neighboring regional populations of Budongo Forest chimpanzees sometimes join together into a party, temporarily move together, and then separate into their respective ranges. On the other hand, the social borders of neighboring populations of the Kasoge chimpanzees are more rigid. Although the basic social organization is similar, there are variations from regional population to regional population and from place to place.

A chimpanzee population of about 100 to 150 animals living in the savanna woodland of the Gombe Stream Reserve, about 170 kilometers north of Kasoge, Western Tanzania, is well known. According to Goodall, only the geographic barrier separates the Gombe Stream chimpanzee population from other chimpanzee populations. All the chimpanzees of the Gombe Stream population travel the entire area of about 77 square kilometers (Goodall, 1965; van Lawick-Goodall, 1968). She says:

> Since chimpanzee parties (groups) in the reserve (Gombe Stream) freely unite from time to time without signs of aggression, they cannot be divided into separate communities (regional populations). It seems likely that only a geographic barrier would constitute a limiting factor on the size of a community, although individuals living at opposite ends of the range might never come into contact (1965, p. 456).

But considering the observations at Kasoge and Budongo, it is possible that two or three loosely separated chimpanzee populations inhabit the Gombe Stream Reserve. For confirmation of such things as social organization, size of a population, and the intergroup relationships, individual identification and long-term observations are essential.

There must be many regional variations in chimpanzee social organization and behavior in the groups that range from the thick forest to the savanna woodland. Basically the thickness of the forest, the seasonal variety and abundance of food, and other ecological factors must be the fundamental factors behind such variation. Population density and other sociological factors influenced by biological and ecological characters are also important for comprehending regional variation in social organization. Possibly, the regional variations in social organization and behavior are dependent upon historically established traditional differences. Young chimpanzees are socialized into their regional traditions—they learn by watching the behavior exhibited by their mother and their elders—and pass this tradition on to succeeding generations.

SOCIAL REGULATION AMONG THE PARTY MEMBERS

When two parties join at a fruit-bearing tree, some kind of greeting behavior occurs between members of both parties, especially between the big adult males. One observation follows. After crying vigorously and beating the buttress of a tree with both hands, a big male climbed the tree with composure. Hearing the greeting cry and beating, a big male from the former occupying party approached the newly arrived big male. The two males embraced and mutally groomed. Then they began to eat. No other chimpanzee performed a similar greeting ceremony. Sometimes a newcomer approaches a former occupant, stretches its hand and both chimpanzees touch each other's hands. Big males may sometimes lessen the psychological friction with a male of another party by touching. Touching may function as a greeting between the leading animals of two parties. Young animals may not feel such friction with others; besides, they cannot perform the complete greeting behavior.

The greeting behavior witnessed when two parties meet seems to be a leader's behavior, for it was seen only between adult males. Other animals in the joining groups neither greeted nor exhibited aggression. These examples show that chimpanzee social tension is solved not by antagonistic but by friendly or associative behaviors, such as grooming, hand-touching, embracing, and other kinds of greeting behaviors. The coexistence of individuals is ensured without antagonism. Young chimpanzee males may learn and identify the behavior patterns of the elders during their developing stage. When they mature and assume the leadership of a temporarily formed party, they can adopt the best behavior in relation to a situation.

LEADERSHIP

The role of permanent leader can hardly exist in a flexible and changing society in which members form only temporary parties that repeatedly divide and regroup. Nine percent of the total number of chimpanzees counted were observed moving alone. Although it may not be essential for females and immature chimpanzees to follow a leader, many times big males, that is, Nyeusi, Mzee, Kubwa, Mrefu, Mzito and Wili, were observed leading their party to a safe place, informing other members of the author's or a villager's presence, or displaying against them in a critical situation. For example, when the author met a party of shy chimpanzees of RP-B, RP-C, or RP-D, one or more big males displayed against him by exposing their whole bodies, vocalizing, and brachiating from branch to branch while other chimpanzees silently hid behind trees and leaves. On another day, when the author was observing the particularly large party of RP-C crossing the grassland between forests, the first animal, a small adult, hurriedly crossed to the other side, and a second, an adult female, returned to the forest. The third and fourth animals, both big males, after watching the author at the forest edge for a minute, jumped back into the forest.

Then the noisy booming clamor suddenly stopped. The author waited another half hour but no further sound of chimpanzees was heard.

Finding the author in the forest, a shy chimpanzee who was not accustomed to him usually went away by itself leaving its party fellows. This behavior was typical in mother parties or in parties lacking big males. Many males may learn how to cope with emergencies with the enemy, and with other critical situations while other animals understand the meaning of the warning behavior of big males and react appropriately. Unfortunately we don't yet know the mechanism(s) whereby the leader's role is learned, but we know that some behavior patterns of big males are actually different from those of others, and that the former leads the party to safety when required.

In the Kasoge chimpanzee population, which is rather compact and separated from neighboring populations, big males may sometimes assume the leadership role in the party (Nishida, 1968). Even in the Gombe Stream chimpanzee population, which is rather loosely scattered, it is the leader who initiates group movement and regulates its speed and direction, although a leader may at any time cease to function (Goodall, 1965).

Although there is no male who always leads and organizes a group of chimpanzees, experienced and strong males can appropriately cope with most situations. So, even in the flexible chimpanzee society where animals temporarily and freely gather and scatter, many individuals tend to gather around big males.

CONCLUSION: THE ROLE OF FLUX IN THE CHIMPANZEE SOCIETY

Many primates live in an organized troop in which all ages and both sexes are included, and in which members always move compactly together as a stable social unit. There is a ranking hierarchy among troop males, although the strictness with which the hierarchy is enforced varies. The ranking relationship is recognized among them and the hierarchy functions to ameliorate conflict. The highest-ranking male or males defend, control, and lead the troop; the strong social bond among members and their safety is maintained.

On the other hand, chimpanzees lack a stable social troop. Even members of a regional population, who are acquainted with each other, rarely move en masse but move in temporarily formed parties that usually consist of less than ten animals. Such parties maintain associative and friendly contact through their rich vocal and behavioral communication. Chimpanzee society ensures the free and independent movement of each individual based on highly developed individuality without the restriction of either territoriality or hierarchy. On the other hand, a chimpanzee enjoys the benefits of group life in that it can avoid the enemy and find fruits with less effort.

Although there is a loose dominant and subordinate relationship among individuals, chimpanzees are rarely placed under the restraint of the ranking hierarchy. The rigidly organized troop characteristic of most primates must be an adaptation for avoiding enemies like man and carnivores and for defense against these enemies. In this context, a group of monkeys is more likely to survive than a single individual. The group provides a social mechanism for survival. Females and young monkeys, especially a female with a baby, must be protected by others. As their food, fruits, nuts, leaves, and some kinds of insects, is scattered in a wide area in the natural habitat, a dominant animal does not control the entire food source, nor does a subordinate animal starve when the former is satiated. An important problem in the rigid hierarchical social organization is that each animal must adjust its movements and behaviors to those of the troop. A rigidly organized troop cannot be maintained when individuals do not subordinate their personal desires for the good of troop unity or solidarity. The flexible social organization of the chimpanzee may be one resolution of this problem. This kind of social organization may be one of the original factors raising individuality to the level of personality. Chimpanzees have not rejected group life, but they have rejected individual uniformity and the pressure of a dominance hierarchy.

That experienced big males can appropriately cope with critical situations as the leader, and that followers appropriately react to the leader's behavior, prove that chimpanzee society is not a simple chaotic gathering but a developed society based on highly developed psychological processes and individuality. The identity of fellow chimpanzees is formed in the mind of those chimpanzees who utilize the same range. The size of the regional population must be restricted by the upper limit of members that an animal can identify and have friendly relations with. Another factor restricting population size must be environmental conditions, that is, the volume and the distribution of food and shelter and the geophysical condition of the habitat. The latter may influence the moving pattern, moving range, and the grouping pattern of each individual and group of individuals. Chimpanzees form regional populations even in continuous habitats such as those found in the Budongo Forest.

When two regional populations come together at the overlapping part of the home ranges of both populations, each chimpanzee, especially a big male, is conscious of the difference between its group and the other population. The strength of the social bond among chimpanzees of a regional population may be copied from the elder's behavior by the young chimpanzees during their development. Variations of social organization from habitat to habitat fundamentally depend on the natural condition of the habitat, on the social background or environment where young animals spend their youth, and on the social tradition or characteristics that young animals learn by copying their elder's behavioral patterns.

A similar pattern of flux in a flexible society is seen in African hunting and gathering peoples, although they have some important social characteristics that chimpanzees lack. Kalahari bushmen of South Africa repeat the pattern of group division and confluence, but their camp is stable and lasts from two to several weeks. Even when a camp breaks up, members of a family or a few families rarely divide. Most of the game and the crop are brought back to camp where they are equally shared with all the camp members (Lee, 1968, 1969; Tanaka, 1969).

Like the Kalahari bushman, Hadza people of East Africa mainly depend on vegetable food. The Hadza male hunts; females and juveniles gather fruits, berries, and roots of plants. They usually do not bring their game and crop back to the camp but eat them where they are. Whatever provisions a man brings back to camp are shared equally among the camp members. Some individuals work eagerly while others live off their labor. This causes people to break camp and separate into smaller groups, but they gather and re-form into a large camp near a water hole during the wet season.

The smallest stable unit of the Mbuti Pigmy of the Congo Forest is not the family, but the individual. Pigmies form hunting parties and they gather in certain places where they can get fruits, honey, or other kinds of food. They move according to the kind and volume of food. Parties form or divide according to the amount of food at a certain place and according to seasonal changes in food supply. Each individual goes to his own desired place and forms a new party of other individuals. This dispersion results in the "political resolution" of psychological and overt contradiction and friction among individuals (Turnbull, 1968). The conscious resolution of social contradiction and psychological friction among individuals as such may not be present in animals.

Acknowledgments

Thanks are due to Professor Junichiro Itani, Director of Kyoto University Africa Primatological Expedition, for his supervision of my work in East Africa, and to Professor Frank E. Poirier of Ohio State University for his kind help in preparing the manuscript.

References

GOODALL, J. "Nest Building Behavior in the Free-ranging Chimpanzee." *Ann. N.Y. Acad. Sci.* 102 (1962), 455–467.

———. "Chimpanzees of the Gombe Stream Reserve." In I. DeVore (ed.), *Primate Behavior: Field Studies of Monkeys and Apes* (New York: Holt, Rinehart and Winston, 1965), 425–473.

ITANI, J. "On the Acquisition and Propagation of a New Food Habit in the Natural Group of the Japanese Monkey at Takasakiyama." *Primates* 1 (1958), 84–98.

———. "From the Society of Non-human Primates to the Human Society." *Kagaku* 37 (1967), 170–174.

——— and A. SUZUKI. "The Social Unit of Chimpanzees." *Primates* 8 (1967), 355–381.

KANO, T. "Distribution of Non-human Primates in Tanzania." (in press)

KAWABE, M. "One Observed Case of Hunting Behavior Among Wild Chimpanzees Living in the Savanna Woodland of Western Tanzania." *Primates* 7 (1966), 393–396.

KAWAI, M. "Newly-acquired Pre-cultural Behavior of the Natural Troop of Japanese Monkeys on Koshima Island." *Primates* 6 (1965), 1–30.

KAWANAKA, K. "Inter-group Relationships of Wild Chimpanzees at the Mahali Mountains." (in press)

KORTLANDT, A. and M. KOOIJ, "Protohominid Behavior in Primates (Preliminary Communication)." *Symp. Zool. Soc. Lond.* 10 (1963), 61–88.

KOYAMA, N. "On Dominance Rank and Kinship of a Wild Japanese Monkey Troop in Arashiyama," *Primates* 6 (1965), 1–30.

LEE, R. B. "The Sociology of !Kung Bushman Trance Performance." In R. Prince (ed.), *Trance and Possession States* (Montreal: McGill University Press, 1968), 35–54.

———. "What Hunters Do for a Living, or How to Make Out on Scarce Resources." In R. B. Lee and I. DeVore (eds.), *Man the Hunter* (Chicago: Aldine, 1968), 30–48.

———. "!Kung Bushman Subsistence—An Input-Output Analysis." In A. P. Vayda (ed.), *Human Ecology—An Anthropological Reader* (New York: Natur. Hist. Pr., 1969).

NISHIDA, T. "A Sociological Study of Solitary Male Monkeys." *Primates* 7 (1966), 141–204.

———. "The Social Group of Wild Chimpanzees in the Mahali Mountains." *Primates* 9 (1968), 167–224.

POIRIER, F. E. "Analysis of a Nilgiri Langur (*Presbytis johnii*) Home Range Change." *Primates* 9 (1968), 29–43.

———. "Behavioral Flexibility and Intertroop Variation Among Nilgiri Langurs (*Prebytis johnii*) of South India." *Folia Primat.* 11 (1969), 119–133.

REYNOLDS, V. and F. REYNOLDS. "Chimpanzees of the Budongo Forest." In I. DeVore (ed.), *Primate Behavior* (New York: Holt, Rinehart and Winston, 1965), 368–424.

SUGIYAMA, Y. "Social Organization of Chimpanzees in the Budongo Forest, Uganda." *Primates* 9 (1968), 225–258.

———. "Comparison of the Primate Societies." *Biol. Sci.* 20 (1968), 113–120.

———. "Social Behavior of Chimpanzees in the Budongo Forest, Uganda." *Primates* 10 (1969), 197–225.

SUZUKI, A. "On the Insect-eating Habits Among Wild Chimpanzees Living in the Savanna Woodland of Western Tanzania." *Primates* 7 (1966), 481–487.

———. "An Ecological Study of Chimpanzees in a Savanna Woodland." *Primates* 10 (1969a), 103–148.

———. "The Society of Meat-eating Chimpanzees." *Shizen* 24 (1969b), 46–56.

TANAKA, J. "The Ecology and Social Structure of Central Kalahari Bushmen." *Kyoto Univ. Afr. Stud.* 3 (1969), 1–26.

TURNBULL, C. M. "The Importance of Flux in Two Hunting Societies." In R. B. Lee and I. DeVore (eds.), *Man the Hunter* (Chicago: Aldine, 1968), 132–137.

VAN LAWICK-GOODALL, J. "Mother-Offspring Relationships in Free-ranging Chimpanzees." In D. Morris (ed.), *Primate Ethology* (London: Weidenfeld & Nicolson, 1967), 287–346.

———. "A Preliminary Report on Expressive Movements and Communication in the Gombe Stream Chimpanzees." In P. Jay (ed.), *Primates: Studies in Adaptation and Variability* (New York: Holt, Rinehart and Winston, 1968), 313–374.

WOODBURN, J. "Stability and Flexibility in Hadza Residential Groupings." In R. B. Lee and I. DeVore (eds.), *Man the Hunter* (Chicago: Aldine, 1968), 103–110.

Junichiro Itani

A Preliminary Essay on the Relationship Between Social Organization and Incest Avoidance in Nonhuman Primates

JAPANESE MACAQUES

During the early period of our research (Itani, 1954), we considered the social unit of wild Japanese macaque troops to be a stable, closed system.[1] However, it is now known to be a semiclosed system whose border is analogous to a selectively permeable membrane. This border is impenetrable for females, that is, it prevents them from leaving the troop; males, on the other hand, can leave or enter a troop. To date our records indicate that a far larger number of males than we first expected leave the troop and begin a wandering life. These animals, called solitary males, can approach another troop and eventually join it. This behavior has been variously interpreted. A solitary male was once considered to be a nonsocial individual, but many cases of males leaving one troop and entering another have now been recorded (Nishida, 1966). It is now clear that a temporary solitary life is a normal characteristic of male Japanese macaques.

This raises the question of why many male monkeys leave the troop where they were born, matured, and even held a stable social status. It is extremely difficult to understand the motivation behind this phenomenon. However, we do know that by leaving his troop a male minimizes the chances of mating with close relatives like his mother, sister, or daughter. When a solitary male joins another troop, we can safely assume that there is no possibility of his mating with his first-born daughter in the new troop for approximately four years, that is, the age at which she reaches sexual maturity. After four years in the second troop there is a possibility of father-daughter mating. Whether or not this is avoided by a second transfer remains unclear.

Our records show many cases of male monkeys who continued to occupy a dominant status in one troop for more than ten years. At Koshima Islet

[1] Editor's note: The Japanese often refer to the troop as the "oikia."

as of August 1952, for example, Kaminari held the status of the most dominant male in the troop and continued to hold it for eighteen years until his death in 1970. Similar cases are reported at Takasakiyama, where Jupiter was the most dominant male from 1953 to 1961, and where Bacchus held a position of high dominance (he was a member of the dominance hierarchy) from 1953 to 1969. It is likely that during such a long tenure mating occurred between these males and their mothers and/or daughters.

Imanishi (1966) reported that a male monkey "may hesitate to mount his mother." It is almost impossible to confirm the degree of inhibition, but this statement does correlate with our findings that the frequency of sexual intercourse between a mother and her son is extremely low. We traditionally assumed that incest is avoided through the psychological recognition of the roles of parent and offspring. However, as far as we know, there is no recognition among nonhuman primates of the father-daughter relationship. Therefore, the most effective means of incest avoidance would be for a male to leave his maternal troop, that is, the troop containing his mother.

Our detailed analyses of the distribution of Japanese macaques show that 75 percent of the troops have neighboring troops within a radius of five kilometers (Kawanaka, in press). Some Japanese macaque troops live close to one another; Kawanaka calls this a "local concentration of troops." This concentration should not, however, be considered a community, that is, a social aggregate higher than the troop. Each troop within a local concentration has a unique relationship with its neighbors. However, bordering troops are not entirely closed to their neighbors, as males may shift from their maternal troop to a neighboring troop. Neighboring troops may be the source of new males for each other, regardless of the daily antagonism they might manifest toward one another.

The process of troop fission has recently been traced by Furuya (1968, 1969) and Koyama (1970). It is necessary to consider troop fission in relation to the extent of troop integration and of troop solidarity. Troop fission is the only means whereby females, who are tightly bound to their nomadic range, can become attached to a new range. When females are involved in a troop fission that produces two heterogeneous troops, males are provided with an opportunity to migrate from their maternal troop to another troop.

The detailed research of Koyama and others during the fission of the Arashiyama troop showed that the troop divided into two halves each consisting of one hundred animals. The sixteen consanguineal groups of the original troop were divided between A-troop and B-troop with no break in the consanguineous relationship. Many of the young males followed their consanguineous relatives; that is, they joined the troop to which their mothers belonged. However, within one year an unusual change occurred: The young males who first belonged to A-troop migrated to B-troop, and those of B-troop joined A-troop (Koyama, 1970). We are unable to construct a satisfactory sociological theory to explain this phenomenon, but we can note that the probability of incest occurring was significantly reduced in the situation where the males left their maternal troop, as opposed to the situation where they did not.

HANUMAN LANGURS

The social organization of Japanese monkeys might reflect a structure common to terrestrial species such as *Macaca* and *Papio,* whose troop size is relatively large and typically includes multiple adult males. However, the arboreal Colobinae have a different social organization. The arboreal colobus are characterized by small, one-adult-male troops. As a case in point, we will look at the Hanuman langurs of south India (Sugiyama, 1965a).

There are two types of Hanuman langur groups: One is a bisexual group usually containing fewer than twenty animals; the other is an all-male group. Bisexual troops are territorial (that is, they maintain a circumscribed area for their exclusive use); but, in contradistinction, the all-male groups appear to lack even a stable nomadic range. Males born into the one-male bisexual troop must leave it before they attain sexual maturity. This would help account for the existence of all-male langur groups, which consist of males unable to join existing bisexual groups or to form bisexual groups of their own. The emigration of a sexually mature male from a bisexual troop largely precludes his mating with his mother and/or sisters, especially since the male seems not to rejoin his maternal group.

The existence of both one-male bisexual troops and all-male groups is a very interesting phenomenon, and although we are not entirely clear as to the functional mechanisms at work, or the exact interrelationships between the two groups, tentative explanations may be advanced. If we assume that the adult male of a bisexual group originally emigrated from his maternal troop and joined an all-male group—and if we further assume that males rarely rejoin their maternal troops—then it logically follows that adult males residing in bisexual groups cannot commit mother-son incest. The probability of father-daughter incest would be 100 percent after the adult male had resided for four years in the bisexual troop, since he has exclusive sexual access to all the troop females, and since four years is the requisite time needed for a female to become sexually mature. (This assumes that no psychological avoidance mechanisms would exist mediating against father-daughter mating.)

Sugiyama (1965b) reports that although the males of an all-male group usually avoid bisexual groups they sometimes attack them. The male of the bisexual troop alone challenges them, but is usually wounded and driven away. Then, one of the males of the all-male group takes his place. Sugiyama states that a male in a bisexual group controls such a group for about three years. Thus incest between himself and his daughter, who was born by the first mating between him and her mother after he assumed the male status of the group, could be completely avoided because he is driven from the group prior to his daughter's reaching sexual maturity.

One further point remains to be made. The following conjecture is based upon the standpoint of its consequences and does not presume to understand the underlying psychological motivations. Upon assuming the position of a bisexual-group male, that male successively attacks infants carried by their mothers. The mothers desert the wounded infants, most of whom die as a result. This behavior signifies the beginning of a new relationship between the new male and the group females. Those infants falling victim to the new male's aggression are all offspring of the former male and the females of the group. Therefore, if the reign of the male in a bisexual group were longer than three years, the possibility exists that some of the infants killed were his offspring, his daughters included. This infanticide plays an important role in selecting against infants born as a consequence of incest. The probability of incest avoidance among Hanuman langurs is greater as a consequence of this phenomenon.

As far as the mechanism of incest avoidance is concerned, it can be said that Hanuman langur society has a much more complete avoidance mechanism than that found among Japanese macaque, although this avoidance is based on the assumption of the sacrifice of all the infants occurring once every few years.

THE ANTHROPOIDS

Among the anthropoids, the gibbons exhibit a unique structure never observed among Cercopithecoids. Among Japanese macaques, and also among Hanuman langurs, females remain in a close relationship with a fixed range all their lives. The unit group of these species is a matrilocal society based on a close tie between the females and their range. However, this principle is not applicable to gibbons. A gibbon group, whose composition is strictly limited to one adult male and one adult female with their young, is neither matrilocal nor patrilocal but is limited to only one generation. A father drives his son from the group when the latter attains sexual maturity; likewise a mother drives her daughter away. This results in the avoidance of mating between a father and daughter and a mother and son (Carpenter, 1940).

The situation obtaining among gorillas is still unclear. However, according to Schaller (1963), who traced one gorilla group for more than a year, a female with an infant appeared from outside and joined a group. Although this is an isolated case, it does suggest that gorilla females do not always have a strong bond with the nomadic range of their native group, or, for that matter, with any group all their lives. It is yet unknown how this female transfer is related to incest avoidance (or if it is), and how general this phenomenon is among gorillas. However, since many of the gorilla groups have only one adult silverback male, and since lone males are encountered, gorilla males, like those of other species, may leave the group before they reach sexual maturity.

Our analysis of incest avoidance among chimpanzees has just commenced. A chimpanzee group is not a one-male group as is the gorilla group (Itani and Suzuki, 1967; Nishida, 1967). The number of males, especially young adult males, is less than that of females. Therefore, we may assume that chimpanzee males also leave their mothers before they reach sexual maturity. The composition of the unit group of chimpanzees indicates that the big males in a unit group are all strangers who appear from the outside.[2] If this is so, then mating between a mother and her son would surely be avoided. As mentioned above, however, several males are included in a unit group, and many observers have noted from their long-term observations that sexual relationships within a chimpanzee group are nearly promiscuous. Since a unit group of gorillas is a one-adult-male group, it is possible that psychological controls reduce mating between a mother and her son; but, on the other hand, among chimpanzees whose unit group contains multiple males and which is also promiscuous it might be impossible to avoid father-daughter mating.

About forty cases of female transfer among chimpanzees have been noted between 1968 and 1970. Female transfer appears to be a common phenomenon among anthropoids. If so, anthropoid society clearly contrasts with that of Cercopithecoid society. I think that the key to the avoidance of father-daughter mating among chimpanzees must lie in the behavior of female transfers. The cases of female transfer confirmed by Nishida and Kawanaka (in press) include those of three females who just reached the age of sexual maturity and of an adult female who recently

[2] Editor's note: See Sugiyama (this volume) for a definition of a chimpanzee social unit.

lost her infant. Kawanaka and I observed the latter returning to her native group. On the other hand, two of the former three females remained in the neighboring unit group whose nomadic range overlapped with that of their native group, and were also observed having sexual relationships with the males therein. If this transfer is general among chimpanzees, then mating between father and daughter might be reduced by it.

Since young adult males, as well as young adult females, leave their native group, mating between brothers and sisters is possible. No material concerning this problem has yet been collected. However, the following supposition can be made: For females, the group in which they mate may be their neighboring group, and the radius of their transfer is never wide. In contrast, male chimpanzees migrate over a range of some ten kilometers or more, and since they lead a wandering life, they might find a group far from their maternal group where they would reside. We sometimes observed lone males wandering the "blank area" eventually outside of the chimpanzee's normal range.

If the phrase "zone range of mating" is applied to the range where young female chimpanzees are able to transfer, it can be considered that its whole is a community of a higher social order than of a unit group of chimpanzees. That is, chimpanzee society is characterized by a unit group formed by males and females who join the group from the outside. Such a group is neither matrilocal nor patrilocal. Social intercourse of some type occurs between chimpanzee groups in a wider geographical area than that bounded by either the nomadic range of a single unit group or a community (zone range of mating), since young males may migrate far beyond this range.

Research to date suggests that local concentrations of chimpanzee groups exist, and further, that each unit group has a fixed nomadic range, although the ranges of different groups often overlap. Each unit group maintains its integrity if contact between adjacent groups occurs, but generally such contact is avoided. Some system of dominance ranking between units in a geographical area is likely.

In conclusion, it is hoped that further diachronic analysis of incest avoidance mechanisms operative in nonhuman primate relationships will prove a useful tool for understanding incest regulation patterns in human societies.[3]

[3] Editor's note: Dr. Itani requested that I see that his article was in proper English. Mr. Jackie Pritchard helped with this, and I am grateful. We trust that we have not misquoted Dr. Itani.

References

CARPENTER, C. R. (1940) "A Field Study in Siam of the Behavior and Social Relations of the Gibbon (*Hylobates lar*)." *Comparative Psychology Monographs* 16: 1–212.

FURUYA, Y. (1968) "On the Fission of Troops of Japanese Monkeys I. Five Fissions and Social Changes between 1955 and 1966 in the Gagyusan Troop." *Primates* 9: 323–250.

———. (1969) "On The Fission of Troops of Japanese Monkeys II. General View of Troop Fission of Japanese Monkeys." *Primates* 10: 47–69.

IMANISHI, K. (1966) *Formation of Human Society.* Tokyo: NHK Books, 1–182.

ITANI, J. (1954) "Japanese Monkeys in Takasakiyama." In K. Imanishi (ed.), *Nihon Dobutsuki* II. Tokyo: Kobunsha, 1–284.

——— and A. SUZUKI (1967) "The Social Unit of Chimpanzees." *Primates* 8: 355–381.

KAWANAKA, K. (in press) "The Inter-troop Relationship of Japanese Monkeys."

KOYAMA, N. (1970) "Changes in Dominance Rank and Division of a Wild Japanese Monkey Troop in Arashiyama." *Primates* 11: 335–390.

NISHIDA, T. (1966) "A Sociological Study of Solitary Male Monkeys." *Primates* 7: 141–204.

———. (1967) "The Social Group of Wild Chimpanzees in the Mahali Mountains." *Primates* 9: 167–224.

——— and K. KAWANAKA (in press) "The Inter-unit Group Relationship of Chimpanzees."

SCHALLER, G. B. *The Mountain Gorilla: Ecology and Behavior.* Chicago: University of Chicago Press, 1963.

SUGIYAMA, Y. (1965a) "Home Range, Mating Season, Male Group and Inter-troop Relations in Hanuman Langurs (*Presbytis entellus*)." *Primates* 6: 73–106.

———. (1965b) "On the Social Change of Hanuman Langurs (*Presbytis entellus*) in Their Natural Condition." *Primates* 6: 382–418.

Gary Mitchell and Edna M. Brandt

Paternal Behavior in Primates*

INTRODUCTION

In some species of New World monkeys, the father assumes the total burden of care for an infant, while the mother's only contacts with it involve nursing and cleaning for short periods of time. In spite of this behavior difference, paternal behavior in nonhuman primates has been largely ignored by primatologists, who have meanwhile extensively and intensively studied maternal behavior. Interestingly enough, students of human behavior have also apparently ignored paternal, in favor of maternal, behavior (Nash, 1965). Nevertheless, the literature of anthropology, psychology, and zoology does contain descriptions of paternal behavior in nonhuman primates, though the studies are often incomplete with regard to observations of the behavior and are often simply anecdotal. It is important for the future, though, to initiate controlled laboratory and field manipulations of interactions between males and infants to attempt to answer the question of what role males play (and what mechanisms and factors are involved) in socializing the young to become viable members of their social groups.

Paternal behavior, as we define it, comprises any contacts between adult or subadult males and juveniles or infants. (These age spans are defined below.) We believe that contacts between juvenile males and infants, though not genuine paternal behaviors, are important in the genesis of paternal behavior and are as important to discuss as true paternal behavior.

There are many kinds of contacts possible between males and the immatures of their groups. Social interactions can range from none at all, to tolerance, friendly interaction, caring for, and even adopting an infant. Other possible interactions are aggressive and sexual. The first sections of the present article contain detailed discussions of the most prominent interactions that occur between males and young in many species of nonhuman primates. The final section of the article includes a list of some of the factors that may affect these interactions.

* An earlier version of this article appeared in *Psychological Bulletin,* 71 (1969), 399–417.

We reserve the term "adult male" for fully mature animals who can reproduce. A subadult male is older than a juvenile but is not yet reproductively mature. Infants are still dependent on their mothers, while juveniles have achieved some measure of independence. The term "immature" refers to both juveniles and infants.

SOCIAL INTERACTIONS

NO INTERACTIONS AT ALL

In some species of primates, there are no contacts between males and young. For instance, in tree shrews (*Tupaia* species) neither the female nor the male shows much parental care. Though the male *T. glis* builds the nest for the young, he does not enter the nest after the young are born until they are weaned, four weeks later (Martin, 1966). In fact, Conaway and Sorenson (1966) observed a captive female *T. longipes* threatening, chasing, and attacking a male as he approached a nest in which another female was giving birth. In *T. belangeri* the female only enters the nest box every forty-eight hours to nurse (Martin, 1966), and the male normally never enters it. With unsuccessful breeding pairs, though, entering the nest box was frequent and was correlated with infrequent suckling and/or cannibalism. It seems that avoiding the nest is normal parental behavior, and that entering the nest, except by the female for suckling, is pathological (Martin, 1968).

In their natural habitat, female bush babies (*Galago crassicaudatus*) with newborn young have never been seen together with their mates (Sauer and Sauer, 1962). Adult males of one species of lemur (*Lemur catta*) ignore newborn infants (Ulmer, 1957).

One observer has reported that among the New World monkeys, as the delivery season in a seminatural group (a group originating from captive animals) of squirrel monkeys (*Saimiri sciureus*) began, the males were actively aggressed by the females and were not allowed in the central part of the troop, which was comprised of mothers and infants. Males are apparently reluctant to enter a group of mothers with infants even to feed. Males have been observed approaching a feeding station, seeing many mothers and infants there, and then going to a different feeding station (DuMond, 1968). Another observer in a field study stated (contrary to the above) that pregnant and nursing female squirrel monkeys did *not* disassociate themselves from the males, but he made no mention of paternal behavior (Thorington, 1968).

When infants are born into a troop of bonnet macaques (*Macaca radiata*), the males may groom the mother but are not allowed to touch the infant. In fact, a mother lets no other animal touch her infant until he is two months old, and she quickly grabs her infant if he is threatened by a dominant male (Simonds, 1965). As the infants get older, though, bonnet males do play with infants (see section on "Level five interactions: grooming and playing").

In another Old World species, the lion-tailed macaque (*M. silenus*), the males will not allow the infants to feed with them and in general have few contacts with young animals (Bertrand, 1969). Male rhesus monkeys (*M. mulatta*) also have little interest in newborns (Rowell *et al.,* 1964) and usually ignore infants in their natural habitat (Kaufmann, 1966). Lindburg (1971) has noted that positive interactions between adult males and infants are indeed infrequent. In an experiment in our laboratory we found that *M. mulatta* males rank lowest in paternal behavior when compared with *M. radiata, M. fascicularis,* and *M. arctoides* (Brandt *et al.,* 1970).

Even though positive interactions between rhesus males and immatures are infrequent, Lindburg (1971) suggests that the males are periodically important as social foci, even though no interactions may be taking place. For example, on one occasion three or four infants of the groups Lindburg studied would sit near an adult male and do nothing else in particular. At other times, infants and males would run off into the trees together. Lindburg found one such pair sitting apart from its group. Perhaps the adult males of species showing no interactions with young nevertheless do contribute to the socialization of the young.

Adult male patas monkeys (*Erythrocebus patas*) show no interest whatsoever in newborn infants, though they will threaten older infants (Hall and Mayer, 1967). Similarly, adult langur males (*Presbytis entellus*) show no interest in a newborn langur. If an adult male frightens a neonate, he is chased away by the mother and by other adult females (Jay, 1963a, 1963b).

TOLERATION

In some primate species, although the males may have no active interactions with young, they do tolerate them. For instance, a male might (1) ignore the young animal while he feeds with him, (2) allow the young animal to precede him when feeding, (3) allow an infant or juvenile to touch or climb on him, (4) not retaliate when a young animal pesters him during copulation, or (5) tolerate a dominance display from an infant or juvenile without retaliation.

In squirrel monkeys (*S. sciureus*) male infants have been observed to display the genitals to the dominant male. This display in this species has sometimes been interpreted as signifying dominance and sometimes frustration and defense. In any case, the adult to whom the infant displays does not aggress the infant (Ploog, 1966). The adult male squirrel monkey is thus usually characterized as indifferent but tolerant toward infants.

All of the infants in a capuchin (*Cebus albifrons*) monkey troop observed by Bernstein (1965) were aggressive at times toward the largest male, but also occasionally approached him while lip smacking to touch his face. When the infants threatened the adult male, he never responded; this lack of response probably indicates a degree of tolerance for infants.

By the time an infant rhesus monkey (*M. mulatta*) is eight weeks old, he approaches, touches, and climbs onto the adult males being groomed by his mother. The adult males usually do not react to the infants, although they are occasionally hostile (Kaufmann, 1966). While Southwick *et al.* (1965) reported that the adult male–infant relationship in rhesus is primarily neutral or indifferent, they did describe some social interactions, discussed later in the present article. Koford (1963a, 1963b) stated that infants of high-ranking rhesus females were able to feed before many of the adult males. One male in a captive group of rhesus monkeys (Bertrand, 1969) showed a preference for one infant, whom he tolerated even when he was in consortship with a female. Three of the adult males were tolerant of all four infants, each of whom could huddle and eat with them.

In other species of Old World monkeys, adult male tolerance of immatures has also been observed. When one adult male crab-eating monkey (*M. fascicularis*) was housed with a familiar female and infant, he generally ignored the infant. (However, he killed the infant of an unfamiliar female [Thompson, 1967].) Gifford (1967) reported that some mature crab-eating males were more tolerant of infants than others and that some males allowed the infants to feed with them after they had kept other animals away. In an unpublished study by Martini and Mitchell of a group of *M. fascicularis* housed outdoors at the National Center for Primate Biology at Davis, California, one of our students noted that one of two infants would approach and sit on both of the two adult males with no fear of being threatened or hurt. The other infant was often threatened by one male and was ignored by the other.

In a study of a troop of Japanese macaques (*M. fuscata*), housed in a two-acre outdoor pen, Alexander (1970) observed mothers of infants of six weeks or less restraining their infants from getting close to the dominant male. As they grew older, however, the infants touched, mouthed, or climbed on him. After the first few days of the infant's life, when the male occasionally showed an inexplicable fear response to him, the male usually ignored the infant.

Ordinarily, the pig-tail adult male monkey (*M. nemestrina*) is aloof from infants; and, although he shows some tolerance for their play, he is for the most part indifferent (Kaufman and Rosenblum, 1967). In both natural and captive conditions, male stump-tail monkeys (*M. arctoides*) were quite tolerant of infants (Bertrand, 1969). They permitted the infants to climb on them and to eat with them. The infants could walk toward a male on a tree branch with impunity, whereas older animals always gave the males the right of way. The males, though, did not themselves encourage contact with the young animals. They did not try to make the infants approach or sit on them and did not watch or retrieve the babies.

Adult and subadult olive and yellow baboons (*Papio anubis, P. cynocephalus*) are completely tolerant of an infant for the first four months when the infant plays near them (Hall and DeVore, 1965). Interactions with the infants are described later in the present article.

Infant Indian langur monkey males (*P. entellus*) over ten months of age approach, mount, and embrace the adult males, while infant females of that age approach and embrace adult females. Thus, in the infant males a weak form of transference from mother to adult males is seen. When a dominant male langur is copulating with an adult female, the pair is frequently surrounded by a group of adult and subadult males and large male juveniles, who dash in circles around them or slap the mounting male. The male is tolerant of these activities (Jay, 1963a, 1963b), though it may be a matter of priorities rather than of tolerance.

Similar behavior at copulation is seen in two-and-one-half-year-old infant chimpanzees (*Pan troglodytes*), who try to push the male off of their mothers. The adult males always tolerate this behavior in infants, but hit out at juveniles (van Lawick-Goodall, 1967).

Ellefson (1966) reported that adult male gibbons (*Hylobates lar*) are completely tolerant of one and one-half year olds but become increasingly intolerant as the young gibbons become older. Juvenile silverbacked gorillas (*Gorilla gorilla*) rarely interact with the adult males, but infants are attracted to them and often play on and around them (Schaller, 1963).

INTERACTIONS

For convenience, we have divided the types of social interactions that are possible between primate males and immatures into eight levels: (1) touching, (2) carrying, (3) approaching, (4) retrieving, (5) grooming or playing, (6) protecting, (7) caring for, and (8) adopting. These levels are further defined where necessary as they are discussed below.

Level One Interactions: Touching. Level one interactions are those involving simply touching the infant. There are only two examples in the literature of this low-level interaction. When a mother with an infant in a captive group of stump-tailed macaques (*M. speciosa*) was first released into the group, the subadult and adult males came toward the infant giving greeting grunts. They inspected, touched, and smelled the infant and manipulated its genitalia (Bertrand, 1969). Adult male olive baboons (*Papio anubis*) approached a newborn infant while lip smacking, touched the infant with hand and mouth, and groomed the mother (Hall and De-Vore, 1965).

Level Two Interactions: Carrying. Level two interactions include contact behaviors such as carrying, holding, cradling, and cuddling the infant and huddling with him. Jolly (1966) saw one instance in which a mother lemur (*Propithecus verrauxi*) allowed a mature male to hold and cuddle her infant while the mother herself fed. Adult male capuchins (*C. albifrons*) occasionally carry infants both dorsally and ventrally (Bernstein, 1965), and infant squirrel monkeys (*S. sciureus*) may on occasion be carried on the male's back (Zuckerman, 1932).

One howler monkey male (*Alouatta palliata*) observed by Bernstein (1964) was frequently associated with a small juvenile. When this male was resting, the juvenile often clung to his ventral surface and was occasionally carried in this position for about one meter.

In a field study of black mangabeys (*Cercocebus albigena johnstoni*), Chalmers (1968) observed one male who associated with the two infants in the troop. The male sat with the two ten-week-old infants, carried them, and helped them over obstacles. These infants were observed to run away from other adults who tried to pick them up. When the infants were three months old, another adult male began carrying and sitting with them. When the infants were four to five months old, a third male was once observed carrying an infant.

Rhesus monkey males (*M. mulatta*) also carry immatures on occasion. A four-year-old castrated male (one of ten castrated males released among normal monkeys in free-ranging colonies) attacked humans who had trapped an infant of his group. When the infant was released, he ran straight to the castrated male, who cradled and carried him for over two hours. Eight months later the same male was seen carrying and grooming another infant who was only three months old (Wilson and Vessey, 1968).

In Lindburg's (1971) rhesus field study, only twice were males observed ventrally carrying immatures. Both males were leader males, and the immatures involved were an infant about four weeks old and a yearling.

When fifteen female juvenile rhesus monkeys at the Wisconsin Primate Laboratory were compared to fifteen male juvenile rhesus monkeys with regard to the quantity and quality of behaviors directed toward a one-month-old infant (not a sibling), the juvenile males directed significantly more hostility toward the infant than did the females. Yet three of the fifteen males did establish ventral contact with the infant within fifteen minutes, and five of the males exhibited other maternal-relevant behaviors, such as lip smacking and grooming (Chamove *et al.,* 1967).

Adult barbary macaques (*M. sylvana*) of both sexes are attracted to infants and show favorable responses to them. The dominant male takes an active part in infant care. According to Lahiri and Southwick (1966), the infant barbary macaque spends an average of 8 percent of his time with the dominant males during the first twelve weeks after birth, and dominant males at times carry the infant.

Not only the dominant males, but *all* the males in a Japanese macaque (*Macaca fuscata*) troop interact with immatures. During the delivery season in the field, adult males hug the infants, take them on their loins, or walk with them (Itani, 1963). These same behaviors were seen in a captive group of Japanese macaques housed outdoors in Oregon (Alexander, 1970). The paternal behavior occurred during the entire year, but peaked during the birth season. A three-year-old juvenile was often carried by the second-ranking male over a two-month period. The male, after inviting the juvenile to ride on his back, carried him. Moreover, he often sat with the juvenile in a ventral-ventral huddle. Similar interactions were seen between the fourth-ranking male and a yearling male, and between this yearling and other low-ranking males. Other males and young behaved similarly but less often.

Hamadryas baboon males (*P. hamadryas*) also show level two interactions. When a female is less than two years old, she transfers the mother's role to a male, so that prior to and during adult life, she will cling to the male's back, or be embraced by him, when she is under stress. This young male later becomes her leader in a one-male group. Because a hamadryas male can protect himself from another aggressive male by holding an infant, the males of this species become more strongly motivated toward maternal behavior than do the males of other species. They may cuddle and carry infants for up to one-half hour (Kummer, 1967). In another species of baboon, the yellow baboon of Rhodesia (*P. cynocephalus*), dorsal and ventral carrying, grooming, and sitting together have been observed in the field (Morgan and Tuttle, 1966).

Level Three Interactions: Approaching. Level three interactions comprise approaching, following, or imitating by either the male or the immature. This level also includes an adult male's adopting an immature's habit. In the laboratory adult male bush babies (*G. senegalensis*) have at times been seen following and imitating infant males (Sauer and Sauer,

1962). Adult male olive baboons (*P. anubis*) approach newborn infants while lip smacking (Hall and DeVore, 1965). Very young (four-to-eight weeks) barbary ape infants (*M. sylvana*) show little overt interest in adult males, whereas older infants (three to four months) take an active interest in getting close to males (Gifford, 1967).

The principal interactions that Lindburg (1971) observed between adult rhesus monkey males (*M. mulatta*) and immatures were (in addition to protecting) sitting beside and traveling beside, or the following of a male by an infant.

Probably the most interesting interactions reported at this level are those of males adopting the habit of an immature. Japanese macaque (*Macaca fuscata*) groups show subcultural propagation of food-eating habits. If an infant has acquired a habit like eating candy, the habit is first imparted to the adult males who care for the infants. If no males look after infants, no males learn the habit (Kawamura, 1963).

Level Four Interactions: Retrieving. When a male takes an infant from his mother, retrieves him from elsewhere, or uses him for protection, level four interactions are taking place. When an observer on one occasion mildly threatened a solitary infant in a seminatural group of squirrel monkeys (*S. sciureus*), a subadult male traveled ten feet to retrieve it. After a few minutes, he broke contact with the infant but still remained nearby. When the observer still stared at the infant, the male retrieved it again and left with it riding on his back (DuMond, 1968). Adult male howler monkeys (*Alouatta palliata*) have also been seen rescuing infants. They may retrieve fallen infants whose mothers have been shot (Collias and Southwick, 1952).

In a field study of black mangabeys (*C. albigena johnstoni*), transfer of an infant from its mother to the single male who frequently associated with infants was seen five times. On no occasion did this transfer involve hostility by the male or unwillingness to part with the infant by the mother (Chalmers, 1968).

Two other species of monkeys have been observed to exhibit level four behaviors. In the unpublished study of Martini and Mitchell mentioned earlier, of a group of *Macaca fascicularis* housed outdoors, this behavior was seen on one occasion by one of our students just after the beta female had groomed the alpha male. The female walked away while the male picked up her infant, held it for a short time, and groomed it for about thirty seconds.

As mentioned previously, hamadryas baboon males (*P. hamadryas*) use infants for protection from other males. When frightened, a two year old or subadult male may grasp an infant and embrace it while turning away from the adult male aggressor. The subadult male may also lower his hindquarters to invite the infant to ride on his back. He then carries the infant (or even a large juvenile) in front of the adult male, whose aggression seems to be inhibited by the sight of the immature. Female two year olds under threat never pick up infants in the way described for males (Kummer, 1967).

Level Five Interactions: Grooming and Playing. Level five interactions —grooming and playing—are much in evidence in the primate literature. Grooming is a friendly and arousal-reducing activity that certainly serves to clean the animals; but, more importantly, it also cements the social bonding within a group.

Probably more is known about rhesus monkey (*Macaca mulatta*) grooming than is known about any other species. Lindburg (in press) found that adult male–juvenile grooming was infrequent in the field. Usually, the male approached and was first mounted and then groomed by the juvenile. Juveniles terminated 52 percent of these bouts. Adult males most often groomed adult females (76 percent of grooming bouts), but groomed yearlings nearly as often as they groomed all other classes combined. On two occasions, males were observed carrying infants ventrally while they groomed them. Adult males were infrequently groomed by anyone other than adult females. Subadult males occasionally groomed immatures, but were never groomed by infants. Koford (1963a, 1963b) reported that juvenile brothers and uncles groom and carry infants occasionally.

In another field study Kaufmann (1967) found that between the birth and mating periods, dominant adult males groomed immatures less than at other times, but immatures did not groom dominant adults less. Among the immatures most often groomed by the adult males were close relatives (siblings, first cousins) and orphans. The males also groomed the young of their close female associates and the young of unusually permissive mothers. These mothers were called permissive because they allowed their young greater than average social freedom; their young became relatively independent of their mothers at an early age. High-ranking males were groomed by immatures more often than they reciprocated, but medium- and low-ranking adult males groomed and were groomed about equally by immatures. Lindburg (in press), on the other hand, found that the grooming of juveniles and infants by adult males showed no relationship to the rank of the male, except that the alpha male never groomed immatures. In one captive group, though, the alpha male sometimes grabbed passing infants, sometimes against their will, to groom them as a female does (Bertrand, 1969).

Adult males of other primate groups have been observed to groom infants: Japanese macaques (*M. fuscata*) (Itani, 1963), yellow baboons (*P. cynocephalus*) (Morgan and Tuttle, 1966), and silverbacked gorillas (*G. gorilla*) (Schaller, 1963). Adult male Japanese macaques (*M. fuscata*) have also been observed to groom the one-to-four-year-old juveniles (Alexander, 1970). The male's object of grooming changes from infant to female according to how his social status rises from young male to leader (Furuya, 1957).

Level five interactions include playing as well as grooming. Primate males often play with immatures. In the laboratory the adult male bush baby (*G. senegalensis*) has been observed to play-jump with an infant male (Sauer and Sauer, 1962). Adult male capuchins (*C. albifrons*) and howlers (*A. palliata*) sometimes play with infants (Bernstein, 1964, 1965). The majority of adult male–infant interactions in squirrel monkeys (*S. sciureus*) involve playful wrestling (Vandenbergh, 1966). Male rhesus monkeys (*M. mulatta*) also play with infants (Southwick *et al.,* 1965). In the field Kaufmann (1967) observed twenty-nine play sessions involving adult rhesus males, and twenty-one of these were with immature animals. Four-year-old subadult males played about twice as frequently as older males.

Four other species of macaque males have been observed to play with immatures. Some mature crab-eating macaque males (*M. fascicularis*) are more tolerant of infants' advances than others and often play intensively with infants (Gifford, 1967). In the field adult male barbary apes (*M. sylvana*) play with juveniles. While one male played with juveniles at least seventy-two times, and another was observed to play less than five times with juveniles, adult males were seldom observed playing with infants (MacRoberts, 1970). Bonnet macaque adult males (*M. radiata*) wrestle with infants over two months old, but they do not threaten in play as do juveniles and infants (Simonds, 1965). Japanese macaque (*M. fuscata*) males also play with juveniles. During the pregnancy season and part of the birth season, the total play responses are dramatically increased between adult males and one-to-four-year-old juveniles. Over three-fourths of the males in the outdoor Oregon troop exhibited this increase (Alexander, 1970).

For several days Kawai (1960) followed a group of *M. fuscata* immatures to an isolated spot and observed them playing around the body of a dead male troop leader who had previously played with them. This observation dramatically indicates the extent of male attractiveness to Japanese monkey young.

The only report of playing between males and immatures in apes refers to the gibbon (*H. lar*). Mild play including chasing, biting, and wrestling may be seen frequently between a gibbon adult male and his offspring once it is six weeks of age (Schaller, 1963, 1965).

Level Six Interactions: Protecting. Level six interaction consists of the protection of the infant from danger by the adult male. Behaviors also included in this class, although seen much more rarely, are (1) baby-sitting and (2) helping an infant over an obstacle or difficult tree-crossing. Adult male howler monkeys (*A. palliata*) are the only males that have been observed to make a bridge with their bodies between trees so that infants can make a difficult crossing (Collias and Southwick, 1952). One male in a black mangabey (*C. a. johnstoni*) group helped infants to negotiate obstacles (Chalmers, 1968).

Hamadryas baboon males (*P. hamadryas*) have been observed apparently baby-sitting. Kummer (1967) reports that the play groups of the one-year-old juvenile baboons are often out of sight of the mother but are formed around a subadult or young adult male. The male does not take part in the play, but a frightened player may run into his arms, and the male then threatens the aggressor. Apparently the male acts as a substitute or baby-sitter for several mothers.

The level six behavior that is observed most often is that of protection, and there are many reports in the literature of males protecting infants. A pair of Zanzibar galagos (*G. senegalensis zanzibaricus*) in a zoo have produced twelve progeny, including three litters of twins. The first twin litter was protected by the male, who would sit in the nest and cover the young with his body in response to danger (Gucwinska and Gucwinski, 1968). Subadult squirrel monkey (*S. sciureus*) males in a seminatural group protected four-to-five-month-old infants, who were at that time becoming increasingly independent of their mothers (DuMond, 1968).

In a captive rhesus group composed of animals caught in the wild, the alpha male did not protect adult or juvenile newcomers, but he did intervene when one of the group's infants was attacked. He also jumped upon animal handlers who tried to catch an infant or juvenile for examination (Bertrand, 1969). Lindburg's (1971) field study of rhesus monkeys gives a number of instances of protection by adult males. While males often attack infants, they also defend them from attacks by others, and will break up disputes between immatures and between other adults and immatures.

On one occasion the dominant male gave pant-threats at a subordinate male who had approached a newborn infant; the second male departed. On another occasion an infant was left behind in a tree when its group fled from children with a dog. The dog kept the infant in the tree, and the monkey group appeared agitated for several minutes. Eventually, as the dominant male chased the children away, a subordinate male went to the infant and walked beside it as it returned to the group.

A dominant male protected two infant crab-eating monkeys (*M. fascicularis*) in a natural group after the infants were captured, made blind, and returned to the group in order to determine if they would survive. On two occasions, older juveniles threatened and grabbed at one of the blind infants, who subsequently screeched. In both instances the mother was not present but the dominant male was nearby. He rushed at the juvenile, chasing him off (Berkson, 1970). In our previously mentioned unpublished study on the same species, we noted that when the infants of the two dominant females played together and one became too rough, the second-ranking male would hold one infant while threatening the other.

MacRoberts (1970) studied free-ranging barbary apes (*M. sylvana*) in Gibraltar. One adult male protected some juveniles in a manner similar to that of adult females. The same adult male also formed what MacRoberts called a close protective relationship with a two-year-old male.

Bertrand (1969) states that in her study, the stump-tailed macaque males (*M. speciosa*), in both natural and captive conditions, were protective toward infants. One zoo-reared male protected only the three infants who actively approached him.

Threats to infants may produce aggressive responses from adult male bonnet monkeys (*M. radiata*). Rosenblum observed a dominant bonnet male, living in a family-plan laboratory situation (with one or more adult females, some juveniles, and the females' offspring), who was terrified of humans and ran to a corner and screamed in fright whenever someone entered the monkey quarters. However, on one occasion the cage door was left open and an infant wandered out; the usually cowardly male then attacked Rosenblum, who luckily escaped (cited in Harlow and Harlow, 1965, p. 332).

During the delivery season of another macaque (*M. fuscata*), adult males of high rank protect one-year-old, and sometimes two-year-old, infants in the same way as do the mothers (Itani, 1963). Alexander (1970) described adult male parental behavior in a group in Oregon very similar to that described by Itani. The dominant males in this troop protected the young. Whenever a neonate was handled by a human being within sight of the troop, the dominant males reacted with intense rage and attempted to attack the human. The subordinate males never showed this reaction. Under most other circumstances, the dominant males ignored human beings.

In yet another macaque species, *M. nemestrina,* the males in some circumstances protect infants. In the laboratory a depressive reaction in five-month-old pig-tailed macaque infants who have been separated from their mothers for two or three days elicits protective behavior from the adult males in their group. The infant sits hunched over, his head down, his facial muscles appearing to relax. He remains motionless except for an occasional "coo" vocalization. When infants showing these behaviors were aggressed by another animal, adult or infant, the adult male of the group threatened the aggressive animal and placed himself physically between the infant and the aggressor. The male, however, in no other way provided comfort to the depressed infants (Kaufman and Rosenblum, 1967).

As mentioned previously, an adult male hamadryas baboon (*P. hamadryas*) baby-sits for a play group of juveniles and threatens the aggressor while an aggressed juvenile runs to his arms. The male is not the exclusive protector of one infant, but is rather a generalized protector, and so the place in his arms is competed for by the members of the play group (Kummer, 1967).

In two other baboon species, the olive and chacma baboons (*P. anubis, P. ursinus*), adult males are quite tolerant and protective toward infants and often carry them on their ventral surfaces. The adult males prevent injuries due to fighting within the group and protect its members against predation (Hall and DeVore, 1965). Even when adult baboon females begin to rebuff an infant after six months of age, the males still remain protective and permissive. At ten months of age, the infant actually spends more time with a play group of peers near the dominant males than it does near its mother. In a crisis a juvenile is more apt to run to an adult male than to his own mother. The adult male's permissive and protective attitude continues until the juvenile is about thirty months old. The juvenile then takes his place in the overall troop dominance hierarchy (DeVore, 1963).

The only report of protection in the vervet monkey (*Cercopithecus aethiops*) comes from Struhsaker (1967b), who observed a juvenile soliciting the aid of one adult male against another adult male.

Chimpanzees (*P. troglodytes*) are at times quite protective of infants. Nissen (cited in Reynolds, 1967, p. 117) saw a full-grown male chimpanzee charge to within a few yards of him to retrieve a juvenile. In one case, when a two-year-old infant male chimpanzee's mother died, he was adopted by his juvenile sister and protected by his adolescent brother (van Lawick-Goodall, 1967).

Level Seven Interactions: Caring for. Level seven interactions involve a male's caring for an infant or juvenile. While caring for an infant may involve all of the separate behaviors already discussed, the level of caring for is unique in that the males involved show these behaviors often and repeatedly, generally toward one specific infant. Indeed, the male may have more to do with a given infant than does its mother. Particularly in the New World monkeys, the male is a very active participant in the socialization of the young. For instance, in a group of titi monkeys (*Callicebus molloch*), which usually consists of an adult pair and one or more young, the male holds and carries the infant virtually all of the time except when it is being nursed. The female often takes the infant for cleaning, but if she is sitting near the male the infant may go to the mother himself. When the mother takes the infant, she licks its genitals (which stimulates urination) and places it on her ventral surface for nursing. When nursing terminates, the infant climbs to the mother's shoulders, from which it moves to the male or is removed by the male (Mason, 1966). Repeated rubbing of the genitoanal region of the male along a branch, a type of marking called anal rubbing, has been shown by adult males while carrying the infant (Moynihan, 1967).

Another New World species, the night or owl monkey (*Aotus trivirgatus*), lives either alone or in monogamous families with one young. Both parents care for the infant, but after he is nine days old he is carried by the adult male at all times except when nursing on the mother (Moynihan, 1964).

As in *Aotus,* there is a monogamous-like bond between an adult male and an adult female in the *Callithrix* marmoset genus (Napier and Napier, 1967). The young (normally twins) are usually carried for about two months by the male parent and are handled by the female only for feeding and cleaning (Stellar, 1960).

In the pygmy marmoset (*Cebuella pygmaea*) paternalistic behavior is also conspicuous. Between birth and six weeks, the one to three infants (usually two) nurse on the mother, but at all other times cling to the male parent. At six weeks they move by themselves and no longer cling (Napier and Napier, 1967).

The golden lion marmosets (*Leontideus* species) also live in small groups and usually have twin births. After the first week the male carries and cares for the infant except when it nurses. The young are independent of the male at four months, though they do not reach adult size until one year (Ditmars, 1933).

As in the species discussed already, tamarins (*Saguinus* species) also live in small family groups. Twin births are usual and the male parent plays the dominant role in bringing up the young (Hampton, 1964; Hampton *et al.,* 1966).

Old World monkey males, in general, do not appear to care for infants as obviously as do the males of the New World species discussed above.

When black mangabey infants (*C. a. johnstoni*) reach the age of four to five months, they spend nearly 70 percent of their time with adult males and only 30 percent with their mothers. Here the males play a large role in caring for the infants. After four to five months of age, though, the proportion of time spent with adult males declines (Chalmers, 1968).

In the macaque species, *M. sylvana,* both adult males in a field group cared for infants. Each male protected one infant and virtually ignored the other infants in the troop. The males carried their infants often, retrieved them, and protected them both from other monkeys and from danger. A mother could take her infant from the male at any time and never objected to the male's attention (MacRoberts, 1970). A field study still in progress by Deag and Crook (Crook, 1970) has revealed that adult *M. sylvana* males in Morocco interact to a remarkable degree with young infants, although apparently not with juveniles. A male carrying a baby may approach or attract other adult and subadult males, and together they may examine, mouth, handle, and groom the infant. A baby moves repeatedly from male to male before returning to nurse on its presumed mother.

In *M. fuscata* there is a close mother-baby relationship for at least ten months. After ten months the adult male leaders and subleaders may take over the care of the infants (Imanishi, 1963). Dominant adult males most often take care of the infants of dominant females, with the result that infants tolerated or cared for by the adult male troop leaders have a good chance of acquiring high social ranks themselves (Kawai, 1965). If an infant acquires a cultural habit like eating candy, it is first imparted to those adult males who take care of infants (Kawamura, 1963).

The third-ranking male, Boris, in the Oregon troop of *M. fuscata* showed parental behavior toward several juveniles (Alexander, 1970). He cared for a female, Gamma, for at least eighteen months, until she was four years old. He groomed, defended, and huddled with her; and his behavior was similar to the protective behavior shown by a mother to her two- or three-year-old offspring. When Gamma became sexually receptive, she and Boris each mated with other monkeys but, curiously, not with each other. Gamma delivered an infant the following birth season, and both before and after her delivery she remained close to Boris, who contacted, groomed, and defended her.

Boris also cared for a male juvenile, Belt, for at least eighteen months, until Belt was two years old. He groomed, contacted, and defended Belt consistently after the latter's mother died. During two years of observation, Boris also intermittently cared for a crippled juvenile female and a two-year-old female with an inattentive mother (Alexander, 1970).

Level Eight Interactions: Adopting. The most maternal-like positive paternal behavior is found in level eight interactions, the cases in which a male adopts an infant or juvenile and fully takes over the role of mother. Obviously, this can happen only after the infant no longer needs to nurse on his mother.

In the hamadryas baboon (*P. hamadryas*) adoptions are not unusual. Kummer and Kurt (1963) reported that when one entire band of baboons was captured and then released into the sleeping place of another band, all of the group members of the former band left the area and abandoned an infant. This infant was immediately adopted by the male of the sleeping party. Kummer (1967) states that motherless infants are invariably adopted by young adult males. DeVore (1963) reports an apparent adoption by a beta male baboon (*P. anubis*) of a sick orphan female infant. The male protected this infant, who became his constant companion.

The final instance of adoption involves a rhesus monkey (*M. mulatta*), castrated at five years of age, who was released into a troop of normal free-ranging rhesus monkeys. The male infant that he adopted was observed riding in both the dorsal and ventral positions on the male, and on one occasion, while he cradled the infant, it attempted to nurse (Wilson and Vessey, 1968).

AGGRESSIVE INTERACTIONS

Not all of the interactions between adult males and immatures are friendly, as our discussion so far might suggest. Adult males are frequently aggressive toward young, and the aggression can range from mild threats through chasing and biting, and can even involve attacking and killing. The following discussion presents examples of all of these behaviors.

Vandenbergh (1963) observed an adult male tree shrew (*T. glis*) kill a young male soon after the young male became mature. Although the killing took place in a large enclosure (fifty feet by fifty feet) and the animal killed was no longer an infant, there is apparently a real danger of adult male infanticide in some captive tree shrews. In fact, cannibalism in captivity is common. While the young are most often eaten by the mother, occasionally another female or a male may do so. Of nineteen births (one to three offspring per litter) that Sorenson and Conaway (1968) observed in *T. montana,* only three animals were not eaten in the first two to three days of life! High-ranking males sometimes fought over a newborn, the winner dragging it off to eat. When a *T. longipes* female was observed giving birth to twins, her first offspring was eaten by the dominant male and her second offspring by another female (Conaway and Sorenson, 1966). Martin (1968), suggesting that reports of tree shrew cannibalism may be instances of abnormal behavior, contends that Conaway and Sorenson's laboratory did not provide the proper conditions for normal parental care.

Male bush babies (*G. senegalensis*) also may eat the young of their groups. Buettner-Janusch (1964) observed a male eating the infants of two pregnancies (one infant each), and believes that it is necessary to remove the male parent from the female before she gives birth in the laboratory. Sauer (1967) removed the male when the infant was one week old and returned him when the infant was one month old. Doyle *et al.* (1967), however, have a naturalistic laboratory setting and have had very good birth success without removing the male.

Natives of the Panama Canal Zone report that adult male howler monkeys (*A. palliata*) attack and kill some of the young males. Moreover, Collias (cited in Collias and Southwick, 1952, p. 143) saw an adult male bite the tail of an infant female howler in half and throw the infant to the ground after the infant fell from its mother and gave a distress call.

When juvenile *Callithrix* monkeys reach maturity, they are not killed but are driven away from the family group by the parent of the same sex (Epple, 1967).

Carpenter (cited in Collias and Southwick, 1952, p. 143) saw many infants killed by adult male rhesus monkeys (*M. mulatta*), and sometimes even by females, when the monkeys were living in an artificially crowded (but free-ranging) population. Although under normal circumstances adult males are indifferent to infants, they may be hostile when an infant approaches (Kaufmann, 1966). If an infant bothers an adult male who is eating, the male may pick the infant up, bite it, and throw it to the ground (Southwick *et al.,* 1965). Bernstein and Draper (1964) released an adult male rhesus into a group of two-and-one-half to three-and-one-half-year-old juveniles that had been living together for two months. Initially, the juveniles avoided the adult male, but eventually some of them approached him, lip smacking and fear grimacing in the process. Though at first the adult male often threatened and chased the juveniles, he himself was occasionally threatened by the larger juveniles. After five days peaceful contact was established with the adult by two small juvenile males.

Studies of the development of laboratory-raised rhesus monkeys have elucidated some interesting points regarding male aggression. Since the mother rhesus monkey treats a male infant differently than she treats a female infant, it is possible that adult males may also react differentially to the two sexes. The male infant is threatened more often and punished earlier and more frequently by its mother, while the female infant is usually restrained, retrieved, and protected (Mitchell, 1968a; Mitchell and Brandt, 1970). Even brutal isolate-reared "motherless-mothers" are more brutal toward male infants than toward female infants (Mitchell, 1968a). Since exposure to an excess of early punishment has been correlated with later hostile behavior (Mitchell *et al.,* 1967), the infant male's characteristic predispositions toward rougher play and rougher infant-directed behaviors are subtly supported by the behaviors of his own mother. Male infants of brutal mothers are among the most hostile animals raised in captivity, and they are particularly hostile toward infants. One such young male savagely attacked a one-month-old infant and bit one of the infant's fingers off (Chamove *et al.,* 1967). Although such behavior is not restricted to the offspring of brutal mothers, there is a higher probability that it will occur in brutally reared animals than in those more gently reared.

Monkeys who have been reared since birth in social isolation also display brutal infant-directed behavior (Arling and Harlow, 1967). Female and male isolates are apparently affected in much the same way; that is, male monkeys who have been raised in isolation display brutal "paternal" behavior, and females brutal "maternal" behavior. Mitchell (1968b) reported that isolate-raised males between the ages of three and five years directed physical aggression toward one-year-old infants, whereas mother-raised males of identical age did not. In addition, these isolate-raised near-adult males threatened infants significantly more often than did control males. In another experiment thirteen-year-old adult male isolates and controls were exposed across a transparent barrier to two different stimulus animals: a normal adult and a one-and-one-half-year-old juvenile female. While the juvenile did not elicit any differences between the isolates and controls that were not also elicited by the adult, the juvenile cooed significantly more frequently in the presence of the isolates and screeched and barked *only* when isolates were present (Brandt *et al.,* unpublished manuscript). Thus, paternalistic behavior in the rhesus monkey is affected by early social deprivation just as is maternal behavior.

In *M. fascicularis* Thompson (1967) observed hostile behaviors between an adult male and an infant in captivity. As mentioned earlier, the male ignored a familiar female and her infant but attacked and killed the infant of an unfamiliar female. Thompson speculated that attacks on infants by males who take over new groups may be advantageous to the male from the point of view of sexual selection, since when a female loses an infant, her estrus is advanced and she becomes potentially receptive to the male. In the unpublished study by Martini and Mitchell on *M. fascicularis* mentioned previously, one of our students noticed that as the dominant male approached the dominant female for sexual relations, he often threatened and bit her infant. Perhaps the infant was not his progeny. In any case, the infant avoided the male even though his mother spent a great deal of time near the male. The subordinate male of the group seldom interacted with or threatened this infant.

Bertrand (1969) noticed that two of three zoo-reared males in a captive group of stump-tailed macaques (*M. arctoides*) occasionally attacked infants, though they usually had few contacts with them. During two years of studying an outdoor troop of *M. fuscata,* six attacks by adult males on neonates were observed, and all were by dominant males. Usually, however, the males ignored the neonates, even showing an inexplicable fear response to an infant only a few days old. After the infant was a few days older, the males occasionally pushed or lightly bit it (Alexander, 1970). Adult patas monkey males (*E. patas*) show no interest in a newborn infant but will threaten older infants, who are always successfully defended by their mothers (Hall and Mayer, 1967).

Sugiyama (1966) observed a natural social change that involved significant interactions between adult male Hanuman langurs (*P. entellus*) and langur infants. A group of male langurs attacked the leader of one troop and expelled him and all other adult males in the troop. Following this, the most dominant male of the attacking group drove all of the infants out of the troop as well. These observations led Sugiyama to artificially exclude the leader male from a one-male troop (most of the troops observed by Sugiyama [1967] were one-male troops). During the first week after the leader was excluded, not a single langur left the troop and no unusual behaviors were observed. On the seventh day, however, the leader of another troop became aware of the excluded male's absence and attacked the unprotected troop. He bit two infants immediately and one of these infants left the troop just after he was attacked. This infant biting continued for three days, and four of the attacked infants in the troop died. A few infants were still alive after three days of biting, but after four days, all of them also disappeared from the troop. The mothers did not give special attention or protection to the inactive and wounded infants. Just after the disappearance of the infants due to the severe attacks by the foreign male, each of the adult females became sexually interested in the male, one after the other, and all copulated with the new leader, who now lived in the center of the troop and was received by all members (these observations strongly support Thompson's [1967] speculations mentioned earlier regarding sexual selection).

The new langur leader also maintained his status in his original troop. He came and went between the two independent troops as the sexual activity permitted. When new infants were born six months later, a female of the original troop borrowed a newborn infant from its natural mother in the new troop. The natural mother followed her infant and adopted mother into the other troop; and, with this impetus, the two troops merged.

Thus, infants as well as adult and subadult males are obstacles to a new leader. The new leader makes new social ties with adult females by advancing their estrus with violent attacks on their infants. The desertion of a bitten and injured infant, though, was made by the mother herself and was not caused directly by the male, who never attempted to take an infant by force. The adult male discerns the infants of his own troop from strange infants and treats the former with permissiveness, while the latter are killed or driven away. Infant biting and the attack upon all adult males by new langur leaders are, according to Sugiyama (1966), characteristic phenomena in the process of social change.

Aggression between males and infants is also seen in chimpanzees (*P. troglodytes*). Although the general rule is paternal permissiveness, occasionally an adult male gently cuffs a juvenile when it begs for food (Goodall, 1965). On twenty-five occasions van Lawick-Goodall (1967) observed an adult male chimpanzee attack a female with an infant in the ventral position, but only one attack was directed toward a female with an infant on her back. Nothing was said about what provoked these attacks, but the sight of an infant on the mother's back may act as an inhibitory signal to the male. Adult chimpanzees tolerate infant teasing while copulating with their mothers, but they hit out at juveniles. On one occasion an infant watched closely and touched both members of a copulating pair—her mother and a male. When her mother screamed and tried to escape from another male who rushed at her, the infant female screamed at the oncoming adult male and hit him. He threatened the infant and then gave up the chase (van Lawick-Goodall, 1967).

Adult male gibbons are completely tolerant of one and one-half year olds but become less tolerant from then on. The eldest offspring, usually about four years old, is harshly treated and is not allowed in or near most food sources until after the adults have finished eating. Adults of both sexes threaten and chase the four year old; and if they catch him, they give him a hit, a hard jerk, or a bite. In this way he is gradually peripheralized from the family, which is a monogamous pair (Ulmer, 1957).

SEXUAL INTERACTIONS

In some cases, primate males are sexually excited by the birth of an infant. Although it is difficult to tell whether the excitation is elicited by the sight of the newborn or by the mother, it is probably the latter. Nevertheless, since the infant may be involved to some extent, some instances of sexual behavior at birth are given below.

In many tree shrews (*Tupaia* species) there is a postpartum heat or estrus. In at least *T. longipes* this estrus often hinders parturition, since the males become sexually excited just prior to and during parturition. They attempt to mount during parturition and for two hours thereafter. After two hours the females are no longer receptive and do not accept the males until after the next birth (Conaway and Sorenson, 1966). A post-partum heat has been observed in *T. belangeri* (Martin, 1968), and in *T. montana* both prepartum and postpartum copulations were seen during one birth (Sorenson and Conaway, 1968).

Comparable behavior has been noted in Old World monkeys and apes. Rowell *et al.* (1964) report that adult male rhesus monkeys (*Macaca mulatta*) occasionally appear sexually excited when they witness the birth of an infant. One orang-utan (*Pongo pygmaeus*) male who was left with his mate during her entire eighth pregnancy was sexually excited by the birth of the baby male. As the adult male tried to copulate, the female tried to avoid him. Because the infant was roughly treated in the struggle, the male was quickly removed from the female (Ulmer, 1957).

FACTORS RELATING TO PATERNAL BEHAVIOR

Several hypotheses concerning the causes and roles of various paternal behaviors in primates have been extracted from this review and are listed subsequently. Each of these ideas is based primarily upon one or at most a few species, and each hypothesis is admittedly quite speculative even at the specific level, to say nothing of extending it to the entire primate order. Nevertheless, the following list of factors can provide a framework for future experimentation on primate paternal behavior.

INFANT VARIABLES

Sex of the Infant. Adult males evidently direct more paternalistic protectiveness toward female infants than toward males and for a longer period of time (Itani, 1963; Kummer, 1967). In stump-tailed macaques only infant males establish lasting relationships with juvenile and subadult males, possibly an extension of the strong male-male attraction that exists in stump-tails (Bertrand, 1969). As the infant male primate matures, he becomes a competitor to the adult male and is forced to the periphery of the troop.

Nash (1965) has reviewed much of the literature relevant to the role of the human father in early experience. The few studies that explore the effects of the paternal relationship on boys and girls separately seem to indicate that absence of the father from the home is more harmful to the later behavior of males than of females. Similarly, Jensen *et al.* (1966)

found that a deprived laboratory environment affected the behavior of male pig-tailed monkey infants more adversely than it did the behavior of female pig-tails. Although the deprivation environment in the latter study involved the absence of more than just the father, its parallel with Nash's data suggests that male primates in general may be more adversely affected than females by the lack of a father or adult male paternal figure in their early development.

Age of the Infant. As the infant matures, less and less parental behavior is directed toward him. This applies to both maternal and paternal behavior. In those primates who live in family groups, the parent of the same sex (or, in the case of the gibbon, both parents) drives the maturing offspring away as it becomes a potential competitor for the parent's mate (Ellefson, 1966).

Orphaned Infants. In several species of primates (for example, barbary and pig-tailed macaques, olive baboon), paternal protective behavior has been seen to increase when a mother has died or has been experimentally separated from her infant. Adult males, under such circumstances, often protect the infant from aggression within the troop and may even adopt the infant and carry it (cf. DeVore, 1963; Kaufman and Rosenblum, 1967; MacRoberts, 1970).

Defective Infants. In both Japanese and crab-eating macaques, paternalistic behavior toward crippled and blind infants has been observed (Alexander, 1970; Berkson, 1970). Berkson (1970) has commented that skeleton samples and field studies have often indicated that primates can live successfully in their groups despite severe defects such as fractured bones, senility, jaw injuries, injuries hindering walking, or congenital bone malformations. Since the condition of a temporarily or permanently defective infant prompts the mother to increase her care activity (Lindburg, 1969a, 1969b; Rumbaugh, 1965), it may also prompt the male to show more paternal behavior.

PATERNAL VARIABLES

Age of Males. As in the Japanese macaque, younger males may interact more with immatures than do older males (Alexander, 1970). In fact, the beginnings of paternal behavior in primates appear before puberty. Rhesus field studies (for example, Koford, 1963b) note that the newborn is often an object of interest to his one-year-old male sibling. In the laboratory the beginnings of rhesus infant-directed behavior are evident at eighteen to thirty months, long before male puberty (Chamove *et al.*, 1967).

A study of the paternal behavior in four species of macaques (*M. arctoides, M. fascicularis, M. radiata, M. mulatta*) (Brandt *et al.*, 1970) revealed that young adult males directed more behavior toward immatures than did fully adult males. A further analysis of the data showed that the interaction scores showed the same trend: Young adults and immatures interacted significantly more than did older adults and immatures. These older adults were also the dominant adults, and perhaps this factor was as important as their ages (see next section).

Dominance of Males. The relationship between the dominance of a male and his paternal behavior is not as yet fully defined. Itani (1963) reported that while the central leaders and subleaders of Japanese macaque troops showed paternal care, the peripheral males rarely did. Rank and frequency of paternal care were not related, though the males whose social positions were changing most rapidly showed paternal care more often. Alexander (1970) reported that while juveniles received as much paternal care from subordinate males as they did from dominant males, infants received 78 percent of their paternal care from dominant males.

Number of Males in a Group. Adult males at all levels of the primate order that live in a one-male group or in a monogamous family group often, but not always (an exception is the patas), display marked behavioral interactions with infants (for example, monkey lemur, marmoset, titi monkey, night monkey, hamadryas baboon, Hanuman langur). There is apparently a transference from the maternal attachment to a paternal attachment. Such one-male groups provide obvious practical advantages to the experimental study of paternal behavior (Kummer, 1967; Sugiyama, 1966).

Familiarity with Mother. An adult male who is familiar with a given adult female will direct more protective attention toward her infant than toward the infants of strange females. Thompson (1967) noted that a crab-eating macaque male tolerated the infant of a familiar female and attacked and killed the infant of a strange female.

Conflict Between Males. Infants may in some cases be a source of conflict between males. Chalmers (1968) saw mangabey males attack other males sitting with or carrying infants, although he never saw aggression by a male toward a mother who was transferring her infant to him. Since the only wounds Chalmers saw during his entire field study were sustained by males who at that time carried infants, he concluded that infants might well have been a major source of conflict between males.

Hormonal Factors. Studies of paternal behavior in Japanese monkeys have suggested that the adult male who shows paternal behavior "behaves as if femininity were strengthened in him" (Itani, 1963, p. 95). Although this is but a statement of the general impression of an observer, Wilson and Vessey (1968) observed marked increases in paternal behavior in two castrated adult male rhesus monkeys. Alexander (1970) believes that the increase in paternal behavior during certain times of the year is at least in part the effect of androgen withdrawal. Although confirmation of this guess is not possible without observing the effects of castration of male monkeys kept in constant conditions, Alexander infers the presence of a yearly androgen cycle. He summarizes the evidence for this from facial and perineal skin color changes, testes size, copulation and masturbation frequencies in his troop of Japanese macaques.

Paternal Affectional System. Harlow and Harlow (1965) have conceived of a paternal affectional system in the rhesus monkey. They see the adult male as a generalized "father" who shows affectional responses to members of the social group but who does not show them differently to his own or other infants.

MATERNAL VARIABLES

Dominance of Mother. The adult female who is dominant is often the female preferred by the leader male. Her infant is tolerated more by the adult males and is actually given food preferences over other infants and, at times, over the males themselves (Koford, 1963a, 1963b).

Consort Relations. When a female is in consort, her yearling has special privileges with the adult male involved. The consort relation therefore provides an opportunity for a considerable amount of paternal protection and tolerance to take place. During the copulatory act between their mother and a male, the young often take liberties with the male that they would rarely, if ever, take outside of the consort situation (cf. Jay, 1963b; van Lawick-Goodall, 1967). The copulating situation appears to be the only exception to a general rule among primates that adult males are intolerant of large juvenile males, but even here it may be a matter of priorities rather than of tolerance.

Time of Year. Itani (1963) reported that the peak of Japenese macaque paternal behavior came during the birth season, and postulated that the behavior might be adaptive in that yearlings might be better protected at a time when their mothers were busy with newborns. While the behavior might indeed be adaptive in that way, the Oregon data on the same species indicate that the males' behavior is not a response to the distress of young when their mothers abandoned them upon the birth of a younger sibling. In fact, the increase in male-young affiliative behavior and play began between the breeding and pregnancy seasons and continued during the pregnancy season, before the birth of any younger siblings. In addition, the increase in paternal behavior was also seen toward two year olds (who would not be as distressed at a new birth as one year olds) and toward orphans (who had lost their mothers over a year previously). Thus, the Oregon data suggest that the prebirth surge of male interest in juveniles is not a response to distress of young. The increase is possibly a reflection of a general seasonal variability of affiliative responses toward *all* members of the troop. Males play more during the pregnancy and birth seasons not only with immatures but with other adults, and do not show a higher proportion of play with immatures than at other seasons of the year. Males, though, do show a greater proportion of affiliative behavior toward juveniles in the pregnancy and birth seasons than during the rest of the year.

Social Change and State of Estrus. When a new leader takes over a troop, the males and infants present a challenge to the new male's becoming fully accepted. It is at this time that primate infanticide on a large scale is observed most frequently in the field (cf. Sugiyama, 1967). The killing of infants has the effect of advancing the estrus of the mohers, making the new leader sexually and ultimately socially acceptable to the females and finally to the rest of the troop.

ONTOGENETIC VARIABLES

Identification. Imanishi hypothesized that the infant male monkey assumed the correct sex role and dominance role by "personal absorption of the personality" of a specific adult male "into his own personality"; that is to say, Imanishi suggested that one of the roles of paternal behavior is to permit identification (in the Freudian sense) to take place for infant males. If the infant identifies with a dominant male, he has a good chance of becoming dominant himself (Imanishi, 1965). This change of identification from the mother to a male was also observed in hamadryas baboons by Kummer (1967), who referred to it as "transference." The concept of identification has proved to be of considerable value in studies of *Homo sapiens;* perhaps it will also be useful in the analysis of primate attachments.

Interest in the Center of the Troop. Those males with a strong interest in the center of the troop can often increase their status by showing interest in young infants of dominant females. Monkeys in the center of a troop often have privileges that young males at the periphery do not have. Being tolerated by adult males rather than aggressed may provide part of the motivation for the development of paternal behavior (cf. Itani, 1963; Kummer, 1967).

Early Experience. Much of the experimental data relating to early experience involve the rhesus monkey. Mother rhesus monkeys punish young male infants more often and earlier than female infants. Punishment, in turn, may promote male assertiveness, since the infants of brutal mothers have been reported to be extremely aggressive (Mitchell *et al.,* 1967), and this early male assertiveness may carry over to the interaction of a juvenile male with a newborn infant (Chamove *et al.,* 1967). Male yearlings are less protective toward newborn siblings than are female yearlings. Hence, differential maternal punishment may partially account for sex differences in infant-directed behavior as well as for differences in peer play. Male one year olds reared by brutal mothers are extremely aggressive toward infants (Chamove *et al.,* 1967). Although male juveniles in general are more hostile toward newborn infants than are female juveniles, protective and permissive behavior toward newborn infants is seen in juvenile males on occasion. This suggests that paternal behavior develops before puberty, just as does maternal behavior, but that paternal behavior develops more slowly. Bertrand (1969) reports that stump-tailed macaque males apparently develop tolerance for infants at a later stage than females do. Her observations of wild-caught and zoo-reared animals led Bertrand to believe that males' behaviors toward infants perhaps depended more on learning than did the behaviors of females.

Abnormally brutal infant-directed behavior can be produced in rhesus males by raising the males in isolation. When the males become sexually mature they are brutal toward infants in much the same way as isolate-reared motherless-mothers are brutal toward their newborn infants. Although this brutality may prove to wane after five years of age as it apparently does in isolate-reared females (Arling *et al.,* 1969), early deprivation still profoundly destroys normal paternalistic patterns just as it does maternal ones.

OTHER VARIABLES

Cultural Propagation. Specific forms of paternal behavior are not always general phenomena seen in all troops or groups of a given species. Many forms of adult male–infant interactions actually may be propagated culturally in much the same way as is sweet-potato washing in the Japanese macaque (cf. Itani, 1963).

Captivity or Crowding. A general effect of crowding in natural, seminatural, or laboratory environments is to increase agonistic behaviors of all kinds. Infanticide by adult males is no exception. Paternal protection decreases and paternal aggression increases under artificially crowded circumstances.

Kinship. If the infant is of the same genealogy as the adult, it may receive more paternal attention. Koford (1963a, 1963b), for example, noted that siblings and even uncles interacted with the young infant rhesus more often than did nonrelatives. Sade (1967) reported that an adolescent male rhesus had his six-year-old young adult brother as his closest companion after his mother died when he was four years old.

Ecology. Jay (1965) has noted that adult male–infant social bonds are strong in terrestrial species and weak or absent in tree-living species. She postulates that this difference might be related to the need to respond quickly to leaders in a ground-living species where fast responses to predators are often primary for survival. The adult male–juvenile bond might thus be especially adaptive, since juveniles are often near the edge of the group. Another piece of evidence for the response-to-predation view is that ground-living species have more rigidly defined and maintained dominance hierarchies than do tree-living species.

Phylogeny of Paternal Behavior. Perhaps future research will disclose a clear phyletic growth in the variety and quantity of paternal behavior. At the present time, however, there is no consistent phyletic trend apparent with regard to either quantity or quality of paternal behavior displayed. Although the lowest prosimian males do not display large amounts of paternal behavior, neither do some of the higher simians (for example, patas). Furthermore, a peak of paternal behavior in the primate order probably occurs among the South American monkeys (titi monkey, night monkey, marmoset), at least in terms of sheer quantity of contact with the infant. The adult males of these species carry the infant at virtually all times except when it is being cleaned or nursed by the mother. Ecology, early experience, and the other factors listed above probably have as much to do with the quantities of paternal behavior displayed as does taxonomic position in the primate order.

DISCUSSION

One of the striking things that emerges from the above review is that the paternal behavior exhibited by primates is extremely variable from species to species. A male primate may kill a young animal, or he may have no contact with it whatsoever, or he may participate to a greater degree than the female in its care. Of course, this variability is hardly surprising given the great variability within the primate order of many other factors, such as ecology and group structure. For example, one would obviously expect to find different degrees of paternal behavior displayed by a male with a monogamous mate and by a male with many females in his group.

In none of the many variations in paternal behavior, though, are there longitudinal, normative developmental data on paternal behavior, nor are experimental data on the factors presented in the previous section abundant. The possibilities presented for research by that list of factors are endless. Imanishi's (1963) notion of identification and Kummer's (1967) concept of transference may be particularly useful ideas in generating laboratory investigations of paternal behavior. For instance, it would be interesting to know what adult behavior would result if an infant of a species with normally little contact between males and young, such as the patas monkey, were made to "identify" with an adult male after transferring its maternal attachment too early. Such a situation might be produced artificially in the laboratory by removing the mother and leaving the infant with an adult male during the early months of life. Conversely, one might investigate the adult behavior that would result from taking an infant of a species with a great deal of adult male–young interaction, such as the titi monkey, and preventing the identification with the male. One might produce this situation in the laboratory by breaking up the monogamous breeding pair and either leaving the female alone with the infant or substituting another female for the male. These investigations would at least partially elucidate the role played by adult males in socializing the young.

The great variability in paternal behavior in primates has of course led investigators to different conclusions depending on the species studied. Opposing views of the role of males in an infant's life are held by recent observers of two different primate species. Alexander (1970) characterizes the Japanese macaque as being intermediate in paternal behavior on the scale between killing infants on sight and fully participating in parental care. Even though Japanese monkey males do display some positive social contact toward immatures, it is a very small amount compared to the enormous quantity of positive social contact that comes to infants from the members of his group other than males. On the other hand, the adult males do account for a very large amount of the agonistic contact experienced by the young, and this contact may be their contribution to the socialization of the young. Alexander postulates that the early punitive relationship could be a mechanism in establishing the authority of the male leaders in a group and could explain both the great stability of dominance structures and the ability of some leaders to retain their status even after their physical power has dwindled. Alexander also suggests that paternal behavior in Japanese monkeys might not have anything at all to do with socializing the young but instead may have evolved simply to protect infants and their lactating mothers.

A contrasting view of paternal behavior is held by Crook (1970), who has observed the extensive positive paternal relationship that exists between adult male barbary macaques and the infants of their groups. Crook believes that the extent of paternal behavior in this species is more than that in any other macaque, and that paternal behavior "probably has major significance for the development of adult social behavior in this species" (1970, p. 132).

One could probably speculate, as Alexander and Crook have both done, on the paternal behavior exhibited in each species of primate. It is premature, though, to speculate on a broad basis on paternal behavior as a whole in primates, since we simply do not know enough about its details. Further study of primate paternal behavior in all its forms will tell us more about evolution of the role of the adult male or "father" in socializing primate infants.

References

ALEXANDER, B. K. "Parental Behavior of Adult Male Japanese Monkeys." *Behaviour,* 36 (1970), 270–285.

ARLING, G. L., and H. F. HARLOW. "Effects of Social Deprivation on Maternal Behavior of Rhesus Monkeys." *J. of Comp. and Physio. Psych.,* 64 (1967), 371–377.

———, G. C. RUPPENTHAL, and G. D. MITCHELL. "Aggressive Behavior of the Eight-year-old Nulliparous Isolate Female Monkey." *Anim. Behav.,* 17 (1969), 109–113.

BERKSON, G. "Defective Infants in a Feral Monkey Group." *Folia Primat.,* 12 (1970), 284–289.

BERNSTEIN, I. S. "A Field Study of the Activities of Howler Monkeys." *Anim. Behav.,* 12 (1964), 92–97.

———. "Activity Patterns in a Cebus Monkey Group." *Folia Primat.,* 3 (1965), 211–244.

———, and W. A. DRAPER. "The Social Behavior of Juvenile Rhesus Monkeys in Groups." *Anim. Behav.,* 12 (1964), 84–91.

BERTRAND, M. "The Behavioral Repertoire of the Stumptail Macaque." *Biblio. Primat. (Basel),* 11 (1969), 1–123.

BRANDT, E. M., R. IRONS, and G. MITCHELL. "Paternalistic Behavior in Four Species of Macaques." *Brain, Behavior and Evolution,* 3 (1970), 415–420.

BUETTNER-JANUSCH, J. "The Breeding of Galagos in Captivity and Some Notes on Their Behavior." *Folia Primat.,* 2 (1964), 93–110.

CHALMERS, N. R. "The Social Behavior of Free-living Mangabeys in Uganda." *Folia Primat.,* 8 (1968), 263–281.

CHAMOVE, A., H. F. HARLOW, and G. D. MITCHELL. "Sex Differences in the Infant-directed Behavior of Preadolescent Rhesus Monkeys." *Child Devel.,* 38 (1967), 329–335.

COLLIAS, N., and C. H. SOUTHWICK. "A Field Study of Population Density and Social Organization in Howling Monkeys," *Proceedings of the American Philosophical Society,* 96 (1952), 143–156.

CONAWAY, C. A., and M. W. SORENSON. "Reproduction in Tree Shrews." *Symposia of the Zoological Society of London,* 15 (1966), 471–492.

CROOK, J. H. "The Socio-ecology of Primates." In J. H. Crook (ed.), *Social Behavior in Birds and Mammals.* New York: Academic Press, 1970.

DEVORE, I. "Mother-Infant Relations in Free-ranging Baboons." In H. L. Rheingold (ed.), *Maternal Behavior in Mammals.* New York: Wiley, 1963.

DITMARS, R. L. "Development of the Silky Marmoset." *Bull. of the N.Y. Zoo. Soc.,* 36 (1933), 175–176.

DOYLE, G. A., A. PELLETIER, and T. BEKKER. "Courtship, Mating, and Parturition in the Lesser Bushbaby (*Galago senegalensis moboli*) Under Seminatural Conditions." *Folia Primat.,* 7 (1967), 169–197.

DUMOND, F. V. "The Squirrel Monkey in a Seminatural Environment." In L. A. Rosenblum and R. W. Cooper (eds.), *The Squirrel Monkey.* New York: Academic Press, 1968.

ELLEFSON, J. O. "Group Size and Behavior in White-handed Gibbons, *Hylobates lar.*" Paper presented at the meeting of American Association of Anthropology, Berkeley, Calif., April 4, 1966.

EPPLE, G. "Vergleichende Untersuchungen uber Sexual und Sozialverhalten der Krallenaffen (*Hapalidae*)." *Folia Primat.,* 7 (1967), 37–65.

FURUYA, Y. "Grooming Behavior in the Wild Japanese Monkey." *Primates,* 1 (1957), 47.

GIFFORD, D. P. "The Expression of Male Interest in the Infant in Five Species of Macaque." *Kroeber Anthro. Soc. Papers,* 36 (1967), 32–40.

GOODALL, J. "Chimpanzees of the Gombe Stream Reserve." In I. DeVore (ed.), *Primate Behavior: Field Studies of Monkeys and Apes.* New York: Holt, Rinehart and Winston, 1965.

GUCWINSKA, H., and A. GUCWINSKI. "Breeding the Zanzibar Galago (*Galago senegalensis zanzibaricus*) at Wroclaw Zoo." *International Zoo Yearbook,* 8 (1968), 111–114.

HALL, K. R. L., and I. DEVORE. "Baboon Social Behavior." In I. DeVore (ed.), *Primate Behavior.* New York: Holt, Rinehart and Winston, 1965.

HALL, K. R. L., and B. MAYER. "Social Interactions in a Group of Captive

Patas Monkeys (*Erythrocebus patas*)." *Folia Primat.*, 5 (1967), 213–236.
HAMPTON, J. K. "Laboratory Requirements and Observations of *Oedipomidas oedipus.*" *Am. J. of Phys. Anthro.*, 22 (1964), 239–244.
———, S. H. HAMPTON, and B. T. LANDWEHR. "Observations on a Successful Breeding Colony of the Marmoset, *Oedipomidas oedipus.*" *Folia Primat.*, 4 (1966), 265–287.
HARLOW, H. F., and M. K. HARLOW. "The Affectional Systems." In A. M. Schrier, H. F. Harlow, and F. Stollnitz (eds.), *Behavior of Nonhuman Primates,* vol. 2. New York: Academic Press, 1965.
IMANISHI, K. "Social Behavior in Japanese Monkeys, *Macaca fuscata.*" In C. H. Southwick (ed.), *Primate Social Behavior.* New York: Van Nostrand, 1963.
———. "Identification: A Process of Socialization in the Subhuman Society of *Macaca fuscata.*" In K. Imanishi and S. Altmann (eds.), *Japanese Monkeys.* Atlanta: Altmann, 1965.
ITANI, J. "Paternal Care in the Wild Japanese Monkeys, *Macaca fuscata.*" In C. H. Southwick (ed.), *Primate Social Behavior.* New York: Van Nostrand, 1963.
JAY, P. "The Indian Langur Monkey (*Presbytis entellus*)." In C. H. Southwick (ed.), *Primate Social Behavior.* New York: Van Nostrand, 1963a.
———. "Mother-Infant Relations in Langurs." In H. L. Rheingold (ed.), *Maternal Behavior in Mammals.* New York: Wiley, 1963b.
———. "Field Studies." In Schrier *et al.* (eds.), *Behavior of Nonhuman Primates,* vol. 2. New York: Academic Press, 1965.
JENSEN, G. D., R. A. BOBBITT, and B. N. GORDON. "Sex Differences in Social Interaction Between Infant Monkeys and Their Mothers." *Recent Advances in Biological Psychiatry,* 9 (1966), 283–293.
JOLLY, A. "Lemur Social Behavior and Primate Intelligence." *Science,* 153 (1966), 501–506.
KAUFMAN, I. C., and L. A. ROSENBLUM. "The Reaction to Separation in Infant Monkeys: Anaclitic Depression and Conservation-withdrawal." *Psychosomatic Medicine,* 24 (1967), 648–675.
KAUFMANN, J. H. "Behavior of Infant Rhesus Monkeys and Their Mothers in a Free-ranging Band." *Zoologica,* 51 (1966), 17–28.
———. "Social Relations of Adult Males in a Free-ranging Band of Rhesus Monkeys." In S. Altmann (ed.), *Social Communications Among Primates.* Chicago: University of Chicago Press, 1967.
KAWAI, M. "A Field Experiment in the Process of Group Formation in the Japanese Monkey (*Macaca fuscata*) and the Releasing of the Group at Ohirayama." *Primates,* 2 (1960), 181–253.
———. "On the System of Social Ranks in a Natural Troop of Japanese Monkeys. II: Ranking Order as Observed Among the Monkeys on and near the Test Box." In K. Imanishi and S. Altmann (eds.), *Japanese Monkeys.* Atlanta: Altmann, 1965.
KAWAMURA, S. "The Process of Subculture Propagation Among Japanese Macaques." In C. H. Southwick (ed.), *Primate Social Behavior.* New York: Van Nostrand, 1963.
KOFORD, C. B. "Group Relations on an Island Colony of Rhesus Monkeys."

In C. H. Southwick (ed.), *Primate Social Behavior.* New York: Van Nostrand, 1963a.

———. "Rank of Mothers and Sons in Bands of Rhesus Monkeys." *Science,* 141 (1963b), 356–357.

KUMMER, H. "Tripartite Relations in Hamadryas Baboons." In S. Altmann (ed.), *Social Communications Among Primates.* Chicago: University of Chicago Press, 1967.

———, and F. KURT. "Social Units of a Free-living Population of Hamadryas Baboons." *Folia Primat.,* 1–2 (1963), 4–18.

LAHIRI, R. K., and C. H. SOUTHWICK. "Parental Care in *Macaca sylvana.*" *Folia Primat.,* 4 (1966), 257–264.

LINDBURG, D. G. "Behavior of Infant Rhesus Monkeys with Thalidomide-induced Malformations: A Pilot Study." *Psychonomic Science,* 15 (1969a), 55–56.

———. "Motor Skills of Infant Rhesus Monkeys with Thalidomide-induced Forelimb Malformations." *Developmental Psychobiology,* 2 (1969b), 184–190.

———. "The Rhesus Monkey in North India: An Ecological and Behavioral Study." In L. A. Rosenblum (ed.), *Primate Behavior: Developments in Field and Laboratory Research,* Vol. II. New York: Academic Press, 1971, 1–106.

———. "Grooming Behavior as a Regulator of Social Interactions in Rhesus Monkeys." In C. R. Carpenter (ed.), *Social Regulatory Mechanisms in Primates,* in press.

MACROBERTS, M. H. "The Social Organization of Barbary Apes (*Macaca sylvana*) on Gibraltar." *Am. J. of Phys. Anthro.,* 33 (1970), 83–100.

MARTIN, R. D. "Tree Shrews: Unique Reproductive Mechanism of Systematic Importance." *Science,* 152 (1966), 1402–1404.

———. "Reproduction and Ontogeny in Tree Shrews (*Tupaia belangeri*) with Reference to Their General Behavior and Taxonomic Relationships." *Zeitschrift für Tierpsychologie,* 25 (1968), 409–495.

MASON, W. A. "Social Organization of the South American Monkey *Callicebus moloch:* A Preliminary Report." *Tulane Studies in Zoology,* 13 (1966), 23–28.

MITCHELL, G. D. "Attachment Differences in Male and Female Infant Monkeys." *Child Development,* 39 (1968a), 612–620.

———. "Persistent Behavior Pathology in Rhesus Monkeys Following Early Social Isolation." *Folia Primat.,* 8 (1968b), 132–147.

———, G. L. ARLING, and G. W. MØLLER. "Long-term Effects of Maternal Punishment on the Behavior of Monkeys." *Psychonomic Science,* 8 (1967), 197–198.

———, and E. M. BRANDT. "Behavioral Differences Related to Experience of Mother and Sex of Infant in the Rhesus Monkey." *Developmental Psychology,* 3 (1970), 149.

MORGAN, M. T., and R. H. TUTTLE. "Intimate Infant–Adult Male Interactions in Rhodesian Baboons (*Papio cynocephalus*)." *Am. J. of Phys. Anthro.,* 25 (1966), 203.

MOYNIHAN, M. "Some Behavior Patterns of Platyrrhine Monkeys: The Night

Monkey (*Aotus trivirgatus*)." *Smithsonian Miscellaneous Collections,* 146 (1964), No. 5.

———. "Comparative Aspects of Communication in New World Primates." In D. Morris (ed.), *Primate Ethology.* Chicago: Aldine, 1967.

NAPIER, J. R., and P. H. NAPIER. *A Handbook of Living Primates.* New York: Academic Press, 1967.

NASH, J. "The Father in Contemporary Culture and Current Psychological Literature." *Child Development,* 36 (1965), 261–297.

PLOOG, D. W. "Biological Basis for Instinct and Behavior Studies on the Development of Social Behavior in Squirrel Monkeys." *Recent Advances in Biological Psychiatry,* 8 (1966), 199–223.

REYNOLDS, V. *The Apes.* New York: Dutton, 1967.

ROWELL, T. E., R. A. HINDE, and Y. SPENCER-BOOTH. " 'Aunt' and Infant Interactions in Captive Rhesus Monkeys." *Anim. Behav.,* 12 (1964), 219–226.

RUMBAUGH, D. M. "Maternal Care in Relation to Infant Behavior in the Squirrel Monkey." *Psychological Reports,* 16 (1965), 171–176.

SADE, D. S. "Determinants of Dominance in a Group of Free-ranging Rhesus Monkeys." In S. A. Altmann (ed.), *Social Communication Among Primates.* Chicago: University of Chicago Press, 1967.

SAUER, E. G. F. "Mother-Infant Relationship in Galagos and the Oral Child-transport Among Primates," *Folia Primat.,* 7 (1967), 127–149.

———, and E. M. SAUER. "The Southwest African Bush Baby of the *Galago senegalensis* Group." *Journal of the South West Africa Scientific Society,* 16 (1962), 5–36.

SCHALLER, G. *The Mountain Gorilla.* Chicago: University of Chicago Press, 1963.

———. "Behavior Comparisons of the Apes." In I. DeVore (ed.), *Primate Behavior: Field Studies of Monkeys and Apes.* New York: Holt, Rinehart and Winston, 1965.

SIMONDS, P. E. "The Bonnet Macaque in South India." In I. DeVore (ed.), *Primate Behavior: Field Studies of Monkeys and Apes.* New York: Holt, Rinehart and Winston, 1965.

SORENSON, M. W., and C. H. CONAWAY. "The Social and Reproductive Behavior of *Tupaia montana* in Captivity." *Journal of Mammalogy,* 49 (1968), 502–512.

SOUTHWICK, C. H., M. A. BEG, and M. R. SIDDIQI. "Rhesus Monkeys in North India." In I. DeVore (ed.), *Primate Behavior: Field Studies of Monkeys and Apes.* New York: Holt, Rinehart and Winston, 1965.

STELLAR, E. "The Marmoset as a Laboratory Animal: Maintenance, General Observations of Behavior, and Simple Learning." *J. of Comp. and Physio. Psych.,* 53 (1960), 1–10.

STRUHSAKER, T. T. "Auditory Communication Among Vervet Monkeys (*Cercopithecus aethiops*)." In S. A. Altmann (ed.), *Social Communication Among Primates.* Chicago: University of Chicago Press, 1967a.

———. "Social Structure Among Vervet Monkeys (*Cercopithecus aethiops*)." *Behavior,* 24 (1967b), 83–119.

SUGIYAMA, Y. "An Artificial Social Change in a Hanuman Langur Troop." *Primates,* 7 (1966), 41–72.

———. "Social Organization of Hanuman Langurs." In S. A. Altmann (ed.), *Social Communication Among Primates.* Chicago: University of Chicago Press, 1967.

THOMPSON, N. S. "Primate Infanticide: A Note and a Request for Information." *Laboratory Primate Newsletter,* 6 (1967), 18–19.

THORINGTON, R. W. "Observations of Squirrel Monkeys in a Colombian Forest." In L. A. Rosenblum and R. W. Cooper (eds.), *The Squirrel Monkey.* New York: Academic Press, 1968, 69–85.

ULMER, F. A. "Breeding of Orangutans," *Zoologischi Garten,* 23 (1957), 57–65.

VANDENBERGH, J. G. "Feeding, Activity, and Social Behavior of the Tree Shrew, *Tupaia glis,* in a Large Outdoor Enclosure." *Folia Primat.,* 1 (1963), 199–207.

———. "Behavioral Observation of an Infant Squirrel Monkey." *Psychological Reports,* 18 (1966), 683–688.

VAN LAWICK-GOODALL, J. "Mother-Offspring Relationships in Free-ranging Chimpanzees." In D. Morris (ed.), *Primate Ethology.* Chicago: Aldine, 1967.

WILSON, A. P., and S. H. VESSEY. "Behavior of Free-ranging Castrated Rhesus Monkeys." *Folia Primat.,* 9 (1968), 1–14.

ZUCKERMAN, S. *The Social Life of Monkeys and Apes.* London: Routledge and Kegan Paul, 1932.

Thomas Rhys Williams

The Socialization Process: A Theoretical Perspective

INTRODUCTION

Little attention has been given to the study of the origin, development, and nature of the socialization process. Most socialization research has been concerned with the consequences of that process for individual behavior.[1] This discussion will attempt to provide a theoretical perspective for an understanding of the socialization process.[2] The discussion begins with a brief review of definitions, types, and some major problems of socialization research. Then a conceptual scheme is developed for use in considering the transcultural and holistic dimensions of the socialization process. The third part of the discussion presents a conjectural model of the trans-temporal dimension of the socialization process. The discussion concludes with some comments concerning study of the socialization process.

[1] The socialization research literature is complex and diverse. Selective bibliographies may be found in Barnouw (1963), Child (1954), Clausen (1968), Cohen (1961, 1964), Goodman (1967), Goslin (1969), Haring (1956), Heinicke and Whiting (1953), Honigmann (1954, 1967), Hoppe, Milton, and Simmel (1970), Hsu (1961), Hunt (1967), Kaplan (1961), Kluckhohn, Murray, and Schneider (1956), Lindesmith and Strauss (1950), Orlansky (1949), Sears, Maccoby, and Levin (1957), Sewell (1952, 1963), and Wallace (1961). The summary and review volumes edited by Clausen (1968) and Goslin (1969) provide nearly comprehensive citations to the literature of socialization research in the decade between 1956 and 1966. A current review of literature in the area of psychological anthropology, which includes references to socialization literature, is to be found in Bourguignon (1972). The references in Haring (1956) and Heinicke and Whiting (1953) provide extensive citations to socialization research in the period between 1940 and 1956. Discussions of selected aspects of socialization research also may be found in Brim (1968), Burton (1968), Etkin (1963), Greenstein (1968), Henry (1960), McNeil (1969), Mead (1963), Shimahara (1970), and Whiting (1968). A review and discussion of research on the socialization process may be found in Williams (1972).

[2] The theoretical approach to be developed here is derived, in part, from recent studies in human behavioral evolution. These include Hallowell's work (1950, 1953, 1954a, 1954b, 1956, 1959, 1960, 1961, 1963, 1965, 1967) and discussions by Campbell (1966), Dobzhansky (1962, 1963), Dobzhansky and Montagu (1947), Etkin (1963), Geertz (1964), Montagu (1962), Roe and Simpson (1958), Spiro (1954), Spuhler (1959), Waddington (1960), and Washburn (1963).

SOCIALIZATION DEFINITIONS AND RESEARCH

The term "socialization" has a long history of varied use (Clausen, 1968, pp. 3, 20–72). Three main views of socialization have emerged in the research of the past forty years. These views tend to correspond to the central orientations of the disciplines of anthropology, psychology, and sociology (LeVine, 1969, p. 505). Anthropologists have used socialization to refer to the intergenerational transmission of culture. Psychologists have used this term to denote the acquisition of controls by infants and children for their basic impulses. Sociologists have employed the term to refer to the training of infants and children for future social performances.

The term "socialization" also has had variable applications in the conduct of research. In anthropology, socialization research generally has been comparative and transcultural, that is, inclusive of data from all cultures. Psychologists have studied socialization as a phenomenon of particular individuals in one culture. Sociologists have been concerned with socialization as a series of events affecting individuals in large social groups in one culture. However, most socialization research in anthropology, psychology, and sociology has been *ahistorical,* that is, it has been concentrated upon present events, acts, or cases.

A review of recent theoretical discussions of socialization in Mead (1963), Clausen (1968), Goslin (1969), and Shimahara (1970) indicates a growing recognition that it is necessary to distinguish between abstract statements—whether these concern intergenerational transmission of culture, learning impulse controls, or learning with implications for future social performances—that are universally true and abstract statements concerning these kinds of events or processes as they occur in specific cultures. In other words, as Mead (1963, p. 185) has noted, whatever their orientations or concerns, students of socialization are beginning to realize that they must make necessary distinctions between statements true in all cultures, or *universals,* and *historical particulars,* or statements true only in one culture.

Following Mead (1963, p. 187), it may be suggested that this distinction can be employed as the basis for definitions of *socialization*—the set of species-wide requirements made on human beings as they learn human culture; and *enculturation*—the process of learning one culture in all its uniqueness and particularity. This theoretical distinction may be illustrated by two statements: socialization—"it would appear then, on the whole, adult *Homo sapiens* has rarely taken it for granted children could or would just naturally learn by spontaneous imitation" (Henry, 1960, p. 304); and, enculturation—"Dusun parents believe a child should stop exhibiting dependent behavior by seven or eight years of age" (Williams, 1969, p. 100).

Dusun children (or Eskimo, Navaho, American, and so on) are *enculturated* into Dusun culture while human children are *socialized* into human culture. This is not to say that Dusun (or Eskimo, Navaho, and so on) children are not human. It is to say, as Mead (1963) does, that a theory of socialization first requires the meticulous transcription and analysis of the details of enculturation in individual cultures, and then a systematic abstraction of statements true of all of the many human enculturation processes. Thus one directly observes the socialization process only as it is manifested in the several enculturation processes.

This theoretical style of defining socialization may be illustrated further by noting some general statements concerning the process of socialization. For instance, among the Arunta (or Aranda), a native society of the central Australian desert, the enculturation process is reported to involve three prominent features: (1) the use of the cultural configurations of kinship and kin groups to provide children with a large number of parent surrogates, (2) a balancing of intense, competing demands by kin groups for children's exclusive affiliations, through appeals to widely shared religious beliefs, such as the concept of spirit beings, spirit descent, and myths of the *alchera,* or "dream time" when all Arunta belonged to the same small local kin group, and (3) use of grandparents in special social roles to provide children with the principal portrayals of Arunta culture, history, and overall cultural *ethos* and *eidos* (for example, cultural form and style; see Kroeber, 1948, pp. 293–294).[3]

When these data of Arunta enculturation are compared with those reported for the enculturation processes in 127 other cultures, the three features appear to be present in approximately 78 percent of cultures of the sample.[4] In approximately 22 percent of the cultures in the sample,

[3] For ethnographic descriptions of Arunta enculturation see Basedow (1925), Berndt and Berndt (1964), Davidson (1926), Elkin (1933, 1954), Elkin and Berndt (1950), Mathews (1907, 1908), Pink (1936), Radcliffe-Brown (1930–1931), Spencer and Gillen (1927, 1938), and Strehlow (1947). This example is drawn from Chapter 7 of my work on the socialization process (Williams, 1972). References to the enculturation processes in the 128 cultures used in the sample are to be found in the Appendix to that text.

[4] Any conclusions drawn from this example must be qualified by both methodological and theoretical exceptions. Some of the cultures used in the sample have histories of culture contact and borrowing and common language and physical origins. This raises "Galton's question" (see Galton, 1889; Driver and Chaney, 1968; Naroll, 1961, 1964; Naroll and D'Andrade, 1963; Greenbaum, 1970; and Ember, 1971), since a transcultural sample ideally should be comprised only of cultures with no history of contact or common linguistic or physical origins. Too, scientific logic requires great care in offering statements that imply that all of the configurations or patterns, trait complexes, and so on, of a culture are involved equally in determining the state of a whole culture or any of its major subsystems. Simply to assert that this is the case begs the question.

TABLE. 128-Culture Sample for Three Features of Enculturation

NEW WORLD CULTURES (39)		PACIFIC CULTURES (31)	OLD WORLD CULTURES (58)		
North America N = 26	*Central and South America* N = 13	*Australia, Indonesia, Borneo, New Guinea, Pacific Islands* N = 31	*Eurasia* N = 15	*Europe and Circum-Mediterranean* N = 11	*Africa (Sub-Saharan)* N = 32
Three features present (23)	Three features present (9)	Three features present (28)	Three features present (9)	Three features present (5)	Three features present (25)
Apache	Camayura	Alor	Ainu	Arab	Ashanti
Arapaho	Jivaro	Arapesh	Andamanese	Lapp	Azande
Cheyenne	Kaingang	Arunta	Baiga	Riffians	Baganda
Comanche	Sherente	Bali	Chenchu	R'wala	Basuto
Eskimo	Siriono	Chamorro	Deoli	Yukaghir	Bemba
Flathead	Tenetehara	Dobuans	Lakher		Bena
Hopi	Wapishana	Dusun	Lepcha		Bushman
Hupa	Witoto	Fijians	Palaung		Chaga
Kaska	Yagua	Ifaulk	Yakut		Chewa
Klamath		Ifugao			Dahomeans
Kutenai		Ilocos			Gusii
Kwakiutl		Kwoma			Kikuyu
Navaho		Lesu			Lamba
Omaha		Malaitans			Masai
Paiute		Malekula			Ngoni
Papago		Manus			Nuer
Sanpoil		Maori			Nyakyusa
Slave		Marquesans			Pygmies
Taos		Murngin			Swazi
Teton-Dakota		Otong-Javanese			Tallensi
Wichita		Pukapukans			Tanala
Winnebago		Samoans			Tiv
Zuñi		Tikopia			Tswana
		Trobriands			Yoruba
		Trukese			Zulu
		Ulithians			
		Wogeo			
		Yungar			

one or more of these features of enculturation either are absent or ethnographic data are unclear. In six cultures (U.S.-American, English, French, German, Israeli, Russian), all three of these features of enculturation are clearly absent. These cultures are predominantly industrial and urban, in contrast to the basically nonindustrialized and nonurbanized character of those cultures possessing all three features of enculturation noted in Arunta culture. The accompanying table lists the names and culture region

NEW WORLD CULTURES (39)		PACIFIC CULTURES (31)	OLD WORLD CULTURES (58)		
North America N = 26	*Central and South America* N = 13	*Australia, Indonesia, Borneo, New Guinea, Pacific Islands* N = 31	*Eurasia* N = 15	*Europe and Circum-Mediterranean* N = 11	*Africa (Sub-Saharan)* N = 32
One or more features absent or data unclear (3)	One or more features absent or data unclear (4)	One or more features absent or data unclear (3)	One or more features absent or data unclear (6)	One or more features absent or data unclear (6)	One or more features absent or data unclear (7)
American	Abipon	Javanese	Balahi	English	Bapedi
Mexican	Ona	Kiwai	Chinese	French	Kongo
Puerto Rican	San Pedro	Nauru	Japanese	Germans	Pondo
	Warrau		Koryak	Israelis	Thonga
			Okinawan	Russians	Tonga
			Rājpūts	Zadruga	Turkana
					Venda

NOTE: For a bibliography of citations to enculturation literature in each of these societies see Appendix I in Williams (1971). See also, Williams (1969, pp. 128–137); Whiting and Child (1953); and Cohen (1964).

locations of the 128 cultures included in this sample.

These comparative enculturation data suggest an hypothesis that as belief, value, and social behavior forms associated with industrialization and urbanization are introduced in and spread through a culture, the number of parent surrogates progressively are reduced, competing demands for loyalty to kin groups increasingly are replaced by the demands for loyalty and affiliation made by new technological institutions (for example, guild, craft, union, firm, and so on), so that traditional social balancing mechanisms are no longer effective, while grandparents come to be increasingly unable to serve as guides for children to a culture's history, ethos, and eidos. Thus following this hypothesis the cultures that have become the most industrial and urban would seem to have turned away from regular inclusion in their enculturation processes of at least three features of enculturation that seem common to many nonindustrial and nonurban cultures.[5]

[5] Whether or not one accepts this hypothesis, such comparative data can be taken to indicate that among the ways humans transmit culture are some features noted in

There are uncertainties among socialization scholars concerning this type of use of comparative data of enculturation involving, (1) the size of samples, (2) reliability of ethnographic reports, and (3) the methods of transcultural analysis. The total number of prehistoric, early historic, and contemporary cultures that can be distinctly identified exceeds 4,000 (Murdock, 1958, p. v). There probably are some 3,000 contemporary cultures. Hence, a sample of only 128 of about 3,000 cultures seems to be a limited basis for drawing general conclusions concerning the socialization process.[6]

The number of descriptive accounts of enculturation processes has increased markedly in the past half century. There now are some fifty enculturation studies that are the products of long-term observations and analysis by trained observers. As recently as 1930 there were only three descriptive accounts of enculturation available for use in comparative research on socialization (Kidd, 1906; Grinnell, 1923; Mead, 1928). By 1950 the number of systematic enculturation descriptions had increased by some twenty new accounts, with Mead leading the way in her research (cf., Mead, 1928, 1930, 1935, 1937, 1949).[7] In the time between 1950 and 1970 the number of systematic enculturation studies has increased substantially.[8] In addition, during this period there also were more than 120 research projects conducted in non-Euro-American cultures on selected topics of enculturation.[9] Thus despite a substantial rate of increase of enculturation studies over the past fifty years, the size of a sample for socialization research remains below 10 percent of the number of all contemporary cultures.

The 128-culture sample used here incorporates the more than fifty descriptions of enculturation (see also Whiting and Child, 1953; Cohen, 1964; Textor, 1967) that have been undertaken by scholars following

---

Arunta culture and also found in comparable form in three-quarters of a limited sample of contemporary cultures.

[6] A sample of 400 cultures, or slightly more than a 10 percent sample of contemporary cultures, seems minimally acceptable to most scholars presently engaged in systematic transcultural study. See, for instance, Murdock (1958, 1967), Naroll (1964, 1968, 1970), and Textor (1967).

[7] Typical studies between 1930 and 1950 included those by Dennis (1940), Granquist (1947, 1950), Raum (1940), Opler (1941), Whiting (1941), Leighton and Kluckhohn (1947), Hogbin (1930, 1931, 1943, 1946), MacGregor (1946), Fortes (1938), and Joseph *et al.* (1949).

[8] See, for instance, the studies of Lewis (1951), Mead and MacGregor (1951), Wilson (1951), Ammar (1954), Richards (1956), Spiro (1958), Lantis (1960), Read (1960, 1968), Richie (1963), Ausubel (1965), and Rabin (1965).

[9] See, for example, Leis (1964), Klopp (1964), De Laguna (1965), Guthrie and Jacobs (1966), and Ainsworth (1967).

generally accepted, "standard" procedures of ethnographic field research. In addition, this sample also includes a substantial number of descriptions of features of enculturation reported by traders, travelers, missionaries, and government officials. The reliability of particular ethnographic accounts, whether concerned with enculturation or some other configuration of culture, continues to be a serious problem in all transcultural analysis. Cultural anthropologists have tended to depend upon their own field research and writing experiences in seeking to portray whole cultures as a basis for their judgments of the reliability of another person's ethnographic accounts. Such judgments can suffer from a lack of first-hand experience in field research and publication, misperception and misreading of data reported, and a tendency to be more or less impressed by knowledge of particular educational and personal experiences of authors. These factors may lead to wide variations in a willingness to accept ethnographic descriptions as valid or useful in comparative studies of culture and society.

Naroll *et al.* (1970) have proposed a set of criteria for inclusion of ethnographic works in what is termed a "standard ethnographic sample." [10] These markers of reliability include whether a reporter lived at least a year among a people and regularly used the local language in interviews and observations without depending on a contact language or use of interpreters.[11] Formulation of such criteria for terming an ethnographic work as "standard" demonstrates a continuing uncertainty over the question of the reliability of data used in comparative studies of culture.[12]

The question of the most appropriate method for comparative study of socialization tends to be divided sharply between an essentially *statistical* approach and one that might be labeled as *contextual*. The statistical method for studying the socialization process makes use of discrete data of

[10] The usual procedures of ethnographic research involve (1) long-term residence, usually more than a year in one or more local communities of one society, (2) learning the local language and then using it regularly in the course of observation and interviews, (3) identification and description of key and ordinary status-role positions and behavior in a society, (4) careful observation and interviewing of persons holding key and ordinary status-role positions, and (5) the use of a clearly defined sample of a population as the basis for study.

[11] These criteria have produced a bibliography of approximately 267 ethnographic references for some 245 cultures. The bibliography is interesting, since by use of these criteria it tends to exclude ethnographies by experienced and professional observers long considered to be standard reference works (for example, E. Evans-Pritchard, *The Nuer*. Oxford: Clarendon Press, 1940), while it includes early nineteenth-century texts (for example, M. Dobrizhoffer, *An Account of the Abipones: An Equestrian People of Paraguay*. London: Murray, 1882).

[12] The critical response to Murdock's text on transcultural analysis of social structure (1949) centered in part upon the "reliability" of the sources used in the sample for research.

enculturation, such as "age at the time of weaning from breast feeding," "cradling practices," and so on, extracted from descriptions of that process in a varying sample of cultures, which then are subjected to analysis through use of statistical techniques and tests of validity, association, and certainty (for example, "*P* values," "phi coefficients," "chi square," "Fisher Exact Test," and so on).[13] The basic procedure involved in this method of comparative socialization research is well illustrated in the innovative work of Whiting and Child (1953), where selected features of culture, such as feeding practices for infants and children and adult magical actions, were abstracted from a sample of seventy-five cultures and treated as "dependent" and "independent" variables to provide opportunity for a statistical analysis of possible causal relationships between child-training practices and personality.

Those persons using a contextual method for comparative study of socialization phenomena are critical of the assumption in statistical studies that the immensely complex integration of culture is amenable to a piecemeal, correlation type of research.[14] The regular dismissal in the statistical method of socialization study of the historic specificity and uniqueness of each culture also is cited by those using the contextual method as a basic error in method, since it leads to a blurring of the two different levels (universal and local) of conceptual abstraction involved in a study of enculturation and a study of socialization. Notable examples of the contextual method of socialization research are found in Cohen's studies of the relationships between social structure and personality (1961), and between initiation rituals and personality (1964). In these studies Cohen used whole ethnographic accounts repeatedly to test specific hypotheses, such as the relationship between adult treatment of child aggression, the fantasies of children, and the transition from childhood to adolescence.[15]

Mead (1963, p. 187) believes that research specialists in socialization have been unwilling to use the contextual method of study because they lack the requisite skills and knowledge gained only through long-term, intensive field study in a number of cultures, which can enable them to clearly differentiate between statements about enculturation and socialization and between an author's presentations of data and his interpretations of data.[16] Mead implies that the statistical method of socialization research often is used as a substitute and rationalization for the painstaking, labori-

[13] For an example of this method of research see Textor (1967).

[14] Mead (1963, pp. 186–187) has aptly described the basic differences between the statistical and contextual approaches to study of the socialization process.

[15] For a discussion of these research procedures see Cohen (1968).

[16] For a further elaboration of the contrasts between the statistical and contextual method in study of the socialization process, see Williams (1972, Chapter 7). See also Spiro (1965).

ous, and detailed analysis required in the contextual method of study.[17]

The contextual method of socialization research was employed in the example given above of some common features of enculturation in a sample of 128 cultures. Before presence or absence of any feature of enculturation was noted, a reading was undertaken of the available ethnographic literature on each culture. Then judgments were made, taking account for each culture of its historical particularity and uniqueness and of the whole configurations of culture in which specific enculturation features occur. However, the possible theoretical value of this example of comparative study of features of enculturation in 128 cultures, that is, of socialization research, still must be qualified by problems of sampling, reliability of data, and method of analysis.

TOWARD A CONCEPTUAL SCHEME FOR THE SOCIALIZATION PROCESS

At present there is no formal conceptual scheme for study of the socialization process. In part, this is related to the major research problems noted above (for example, sampling, reliability, method). It also derives from the fact that most often socialization research has been viewed as an interdisciplinary interest and not as concerned with a sharply definable process (Clausen, 1968, p. 5).

However, using the 128-culture sample discussed above it is possible to abstract a preliminary socialization process conceptual scheme from existing data of enculturation. A socialization process conceptual scheme should reflect knowledge inclusive of all human cultures and be founded upon data of whole cultures. Thus the development of a preliminary conceptual scheme can provide for *transcultural* and *holistic* statements concerning the nature of the socialization process. The preliminary conceptual scheme offered here is not an explanation *obscurum per obscurus;* that is, this scheme is not an effort to move discussion away from an unknown process to an even more vague conceptual design, to avoid hard questions concerning the origin, development, and nature of socialization.[18] Rather, this discussion is intended to summarize, in a formal manner, the outline of what may be taken now to be the structure and functions, that is, the nature of the contemporary human process of socialization.

A scientific conceptual scheme may be defined as a set of increasingly more abstract statements of specific empirical references that note the de-

[17] For bibliographies of transcultural studies using an essentially "statistical method," see O'Leary (1969) and Naroll (1970).

[18] This phase may be glossed as the act of explaining an unknown phenomenon through use of a still more unknown phenomenon.

terminate conditions under which objects or events are related among themselves. The ideal scientific conceptual scheme is one in which the categories summed up by general statements are limited in number and type. Conceptual schemes may have several levels of generalization. A conceptual scheme provides a formal intellectual tool for making order from great masses of data, which seem on initial observation and study to be disparate and unrelated.

Scientific conceptual schemes are dynamic entities. Once developed, conceptual schemes tend to encourage seeking of new data to fit existing categories. Sometimes new data cannot be fitted to existing conceptual categories. This situation can lead to a reworking and redefinition of a scheme (cf., Kuhn, 1966). One example of a scientific conceptual scheme that is dynamic in nature is the periodic table of the elements in modern chemistry. When first developed by Dmitri Mendeleev between 1869 and 1871, the periodic table contained sixty-three elements classified by their atomic weights. Today, this scheme contains forty new elements. The contemporary periodic table also includes data concerning the crystal structure, density, and boiling and melting points for each element. All the new elements were added by discoveries made when investigators noted discrepancies in the existing chemistry conceptual scheme. Thus, once developed, conceptual schemes can be used to predict the presence and appearance of phenomena, or to suggest relationships among phenomena, not previously noted or thought possible.

Some physical and biological sciences have developed highly precise, determinate, and general conceptual schemes. On the other hand, social science fields such as anthropology, sociology, and psychology at present have a variety of conceptual schemes that tend to be limited in precision and generality. The aim of all scientific activity is to construct conceptual schemes that can be made increasingly more precise, determinate, and general for use in an explanation, ordering and predicting interrelations between phenomena.

The general theoretical approach to be used here in outlining a preliminary conceptual scheme for the socialization process is derived from the structural and functional theoretical tradition of modern anthropology and sociology. This tradition and its scientific uses are discussed in detail in Merton (1957), Levy (1952), Firth (1956), and Hallowell (1956). The concept of structure will be used here to refer to the arrangement and interrelation of parts as dominated by the general character of the conceptual whole. In the discussion to follow the conceptual whole to be considered is the socialization process. The parts comprising that whole are the patterns and configurations of culture especially concerned with the transmission of human culture. These patterns and configurations may be said to be functionally interrelated in that they exist as the result of the operation of the structure of the socialization process through the time in which culture has existed as a definable and recognizable entity.

Conceptually, the socialization process may be viewed as consisting of a number of functionally interrelated parts. The structural parts of that process may be identified tentatively as (1) *personal agents,* (2) *impersonal agents,* (3) *formal social groups,* (4) *informal social groups,* (5) *a socialization conceptual system,* (6) *total human biological equipment,* and (7) *internal conditions of culture.* The functions of these specific parts of the socialization process may be identified tentatively as (a) *explicit functioning,* (b) *implicit functioning,* (c) *deutero functioning,* (d) *generative functioning,* (e) *anticipatory functioning,* (f) *developmental functioning,* and (g) *linguistic functioning.*

STRUCTURE OF THE SOCIALIZATION PROCESS

In brief outline form, the details of the structure of the socialization process are described in the following paragraphs:

(1) *Personal Agents.* In each one of the 128 cultures used as the basis for development of this conceptual scheme, the process of enculturation involves transmission of culture by adults, as well as by a child's playmates and age-mates. In the initial portion of the process of socialization, a human infant primarily is in direct contact with parents and parent surrogates. These contacts gradually broaden to include other family members and subsequently come to include play-group members and friends. Between the ages of six and ten, depending upon local cultural tradition, children also begin to have direct personal contacts with adults possessing special skills, knowledge, and abilities. In cultures where such contacts are organized systematically, transmission of particular cultural knowledge may be offered in special settings ("schools") by trained adults ("teachers"). In cultures where such contacts are not systematized, adults holding special status-role positions ("priest," "shaman," "craftsman," "hunter," and so on) also may serve as personal agents in the process of cultural transmission.

It should be noted that research concerned with the consequences of the socialization process has emphasized the importance of personal agents of cultural transmission. For instance, Honigmann (1969, p. 189) identifies "human links" (for example, personal agents) as the most vital aspect in the consequences of the socialization process. Honigmann (1967, p. 172) also notes that the interactions between mothers and their infants are the most crucial for the consequences of being socialized. Parsons (1951); Parsons and Shils, 1951) also has offered similar conclusions, particularly in his work with Bales (1955) where the "mother-infant dyad" is said to be of primary importance in the consequences for the individual being socialized. Among others, Homans (1950, 1961) also has advanced the idea that the direct contacts of infants with mothers, as personal agents of socialization, are of paramount importance in transmission of culture.

However, there is no substantive evidence to demonstrate that only one of the different classes of personal agents of cultural transmission has more importance and is more critical to the existence and operation of the socialization process than any other of the classes of personal agents. The structural-functional conceptual scheme proposed here has no requirements that some categories, or parts, be chosen over others as being of primary importance.

(2) *Impersonal Agents.* Human life regularly involves contacts with many kinds of natural phenomena such as wind, sound, clouds, water, sun, stars, plants, animals, and so on. Although these are distributed differentially and have variable local manifestations, and although contacts may vary from proximate to distant, all human groups appear to utilize them in some form or another in the transmission of culture. All groups in the 128-culture sample anthropomorphize and reify some natural phenomena in songs, folk tales, proverbial sayings, riddles, and so on, which are told and sung by personal agents of socialization to infants and children. The most highly personalized of these natural phenomena are meaningful parts of the cultural transmission process. For example, the anthropomorphization and reification of the sun, moon, stars, and certain animals dominant in a local ecology (for example, "moon mother," "sun father," "wolf brother") often take a form where young children may come to converse directly with these natural elements as they would with a parent or a playmate. Such interactions also may occur with objects, either natural or manmade. In more technologically complex cultures, objects made by man, such as books, radios, television sets, motion pictures, computers, and so on, have come to occupy a similar place in the cultural transmission process.

In cultures where children regularly converse with natural elements, the human capacity for reflexive thought (cf., Williams, 1959), symbolizing, and language provides the opportunity for children to conduct "internal" conversations on behalf of, and in the role of, the inanimate and mute impersonal agents addressed; that is, children have and use the capacities for language, symbolization, and reflexivity to complete, on behalf of the impersonal agent, the "other side" of a conversation concerning their behavior or their plans for action. Thus it is the use of these human capacities for animating the inanimate and making personable the impersonal that bring such agents fully into the cultural transmission process. In cultures that possess the kinds of technology that actually can "talk back" to children, whether in the forms of a silent exchange (for example, books) or in the forms that really are verbal and often visual as well (radio, television, motion pictures, and so on), children can readily acquire great amounts of cultural information.

(3) *Formal Social Groups.* Each one of the 128 cultures in the research sample possesses formally structured social groupings that are involved actively in the process of cultural transmission. In small societies such groups usually number from two to twenty persons. These groups are characterized by a face-to-face interaction between members and children. In large societies formal social groups may comprise hundreds, or thousands, of members, and become characterized by face-to-face interactions occurring primarily between specially designated functionaries and representatives (for example, priest, president) of the group and maturing children. Formal social groups exist over long periods of time and have at least four distinctive features, including (1) a clear identification of group aims, (2) specific controls for behavior of group members, (3) a high valuation of intragroup relationships, and (4) the expectation of inclusive knowledge by each member, or a representative of the group, concerning the life history of other members.

Examples of some formal social groups, from both small and large societies, would be the family (nuclear or any of its extended forms), churches, clubs, fraternities, clans, schools and the kibbutz. The expectations for behavior held by members of formal social groups usually correspond with expectations generally known in a culture. Thus when young members are asked to conform to formal group aims, they are in effect being subjected to a continuation of the local process of cultural transmission. Formal social groups exercise a different type of authority from that employed by personal agents in cultural transmission. If new members will not conform regularly to formal group goals and standards of behavior, they face the threat of expulsion from the group. Usually parents and parent-surrogates cannot expel or banish infants and children from their presence in any formal way. As a consequence, formal social groups may take on an active role in the cultural transmission process when parents or parent-surrogates turn to the group for use of its disciplinary powers.

(4) *Informal Social Groups.* Each one of the 128 cultures of the sample contains informally structured social groups, that is, groups with

formally stated aims or goals, but without controls for behavior of its members, with a low level of value on intragroup behavior, and for inclusive knowledge of the lives of group members. Some examples of informal social groups would be play groups, neighborhoods, work teams, feasting "societies," age groups, sex groups, economic interest groups, craft groups, "gangs," blood brotherhoods and other such groupings organized on a nonkinship basis.

Informal social groups typically are organized only for brief periods for the accomplishment of limited aims. Personal relations within such groupings are marked by a "personal impersonality," with members typically expressing little interest in the behavior of other members outside their activities together. Such groups usually operate only with those norms and standards of personal behavior typical of the culture in which the group exists.

Individuals become members of informal social groups because they share in the special interests of these groups. Thus when young children become part of an informal group, it is most often because they have sought it out for its special aims (to play, to learn a craft, and so on). Once part of such an informal group, however, children have a special and "unofficial" opportunity to learn the norms of a local culture, as these are used regularly by older members in the course of group action toward furtherance of their special interests. Thus informal social groups play a part in the cultural transmission process, as they serve the purpose of providing repeated access to the context of usual ideas and behavior standards transmitted and available in each culture.

(5) *A Socialization Conceptual System.* Each of the 128 cultures in the sample possesses a body of culturally transmitted ideas concerning the cultural learning capacities, styles, and potentials of infants and children and regarding the proper social positions and roles for infants and children in everyday life. Each culture also has beliefs about the nature of infant and child life, as well as a distinct set of ideas concerning the whole socialization process. Thus although the forms and meanings of concepts of cultural transmission vary widely among cultures, all groups of the 128-culture sample have and regularly use such concepts in the process of cultural transmission.

The existence of a socialization conceptual scheme as a distinct configuration of each local culture means adults have and know a plan and use assumptions concerning cultural transmission that can be referred to in the course of the long-term operation of that process. Also, children can learn details of this plan as they experience it and through a type of *autokinetic effect* come to apply it to themselves as a guide for their own cultural learnings. This effect is most directly visible when children subject themselves to the norms and standards of adults, even though no punishment or censure could be applied to them in the absence of adult awareness of the acts undertaken alone. And, the existence of a socialization conceptual scheme can provide for a "feedback" or a regular exchange of information between children and adults concerning the process of cultural transmission (cf., Mead, 1970).

(6) *Total Human Biological Equipment.* The phrase "total human biological equipment" refers to the fact that first, in the course of general biological evolution, and then later, in the interactions between hominid biology and social and cultural experiences, there came into existence a complete biosystem that prepares the human infant to be socialized. La Barre (1954) has provided one of the best accounts of the emergence and operations of this biosystem for socialization. Honigmann (1967, p. 177) recently has described this total system and has termed it "an active propensity for socialization," while Hallowell (1953, p. 612) has called it an "inherent organic potential" for socialization (see also, Chomsky, 1968, and Bruner, 1956, 1966).

This does not mean that since *Homo sapiens* is born with a biosystem that prepares him to be socialized, the socialization process is essentially biological in nature. It does mean that *Homo sapiens* infants are usually born possessed of species-characteristic behavior forms, reflexes, drives, and capacities that serve together as organic bases in learning of culture. However, in the absence of cultural transmission, no amount of biologically transmitted readiness, or active biological propensities on the part of the infant, can bring the socialization process into operation. The converse is also true, for no amount of contact with the socialization process really will overcome major deficiencies in the total biological system of an infant, to lead to a full expression of what may be termed "humanness."

(7) *Internal Conditions of Culture.* There was a period in hominid evolutionary development when such populations lived more in a *natural* than a *cultural* ecology. As a cultural system developed gradually and over a long time period, it came to be transmitted to and acquired by hominid young. This process probably was very fitful and incomplete for hundreds of generations, as hominids continued to live in and under conditions of a partly natural and a partly cultural ecology. However, as a cultural system came into existence and slowly developed, the system began to interact more frequently with hominid biology. In turn, this led to a powerful mutual feedback system being activated between hominid biology and cultural experience. Natural evolutionary selection pressures that

solely had affected hominid biology became altered, in time, because of the interactions between that biology and the evolving hominid cultural system. This reciprocal feedback process, from hominid biology to culture and culture to biology, was a new phenomenon in nature, which led to rapid acceleration of events and sequences in both the hominid biological and cultural realms; that is, time intervals between significant and major biological and cultural changes probably began to contract sharply as the mutual feedback between hominid biology and culture operated to produce a vigorous interaction synthesis of both phenomena.

Perhaps more important than a "speeding up" of biological and cultural changes was the fact that the mutual feedback process between hominid biology and culture came to have an identifiable locus, or place, in nature. The course of development of that locus is at present only vaguely known and can be misunderstood easily without a careful exposition of the topic. This topic will be discussed in greater detail in the section on "Origin and Development of the Socialization Process" in this paper.

FUNCTIONS OF THE SOCIALIZATION PROCESS

In brief outline form, details of the functions of the socialization process are described in the following paragraphs.

(A) *Explicit Functioning.* Each one of the 128 cultures in the sample used to develop a conceptual scheme for the socialization process enculturates infants and children according to specific conceptions of the distinctive features of culture. The phrase "explicit functioning" refers to the personal involvement of adults in transmission of the features of culture believed to be distinctive and important. For example, Bateson and Mead (1942; Mead, 1956) report extreme concern by adult Balinese for maintaining an unimpaired body surface. This adult Balinese preoccupation is transmitted through direct comment by adults prohibiting children with open wounds or new scars from participating in religious activities. Such expressions are intentional and overt acts. This concern is part of a set of complexly interrelated cultural patterns, including wound treatment, body cleanliness, mortuary rites, and special religious rituals that the Balinese view as vital and necessary in their lives. Thus explicit functioning of the socialization process involves personal agents in a direct transmission of cultural features believed to be of significance in a culture.

(B) *Implicit Functioning.* An illustration of the implicit functioning of the socialization process also can be taken from Bateson and Mead's (1942; Mead, 1956) description of Balinese concerns for maintaining an unimpaired body surface. Balinese children learn of this adult preoccupation in at least three ways in addition to the direct and active admonitions of personal agents of socialization. These ways are (1) through the repeated expressions of concern by adults to one another concerning the time and manner of wound healing, (2) in the ways children witness and take a peripheral part in a series of mortuary rites, which are repeated over many years for each deceased individual and which involve symbolic attempts by adults to eliminate the corpse and all the objects associated with it, and (3) in the ways adults exert regular and continuing efforts to make a corpse whole and integral again through manufacture of surrogates for use in mortuary rites.

These acts by Balinese adults involve an implicit transmission to the children of the basic importance of body integrity. There is no intentional effort to relate these cultural features to children, for they are considered to be adult concerns and not educationally relevant. Thus Balinese children also are enculturated through learning about their culture in ways that do not involve explicit, overt action by adults. DuBois (1956) has termed this functioning of the socialization process as "absorptive," that is, as a presentation of cultural patterns that are so consistently observed in adults that children acquire them through a "kind of psychic osmosis" (DuBois, 1956, p. 242).

(C) *Deutero Functioning.* In the course of considering different styles of cultural transmission, Bateson (1942) and Bateson and Reusch (1951, pp. 215–516) have noted the fact that children also may learn culture in the acts of learning culture; that is, children may acquire culture in a deutero, or secondary, manner without either explicit or implicit socialization process functions being involved. Bateson (1942) illustrated the deutero function in cultural transmission by noting the ways in which some American schoolteachers and administrators provide conditions for pupils to learn specific American values, such as tolerance, equality, freedom, justice, fairness, and so on. Students soon learn, from the contexts of such teaching, that their adult mentors believe in these values on two levels. One level is the ideal appearance of the values taught. A second level consists of the real opinions of adults involved in teaching values. When American schoolchildren learn the appearance of specific values in a class setting in which no racially, socially, or culturally "different" children are regularly present because of their being excluded by school officials and teachers or because of cultural and social patterns of inequality in income, housing, and parental educational achievements, then children have transmitted to them, in a deutero fashion, the fact that American values apply to certain kinds of Americans and not to all others.

It is useful to note that although deutero and implicit functions seem to be conceptually similar, they are, in fact, separate aspects of the socialization process. One way of understanding the distinction between the concepts would be to note that in implicit functioning, children acquire culture as a result of some directly observed action by adults, even though such actions are not intended by adults to convey significant and important features of culture to children. In contrast, in deutero functioning, children acquire culture essentially from the contexts of their learning and not from the particular overt acts of adults in those contexts. In other words, the distinction between implicit and deutero functions of the socialization process resolves to the facts of cultural transmission through specific adult acts, although these are unintentional, and through the contexts in which cultural learning occurs.

(D) *Generative Functioning.* Until recently the socialization process was portrayed largely as a series of external forces imposed upon infants and children as they matured in a cultural milieu. Allport (1955, p. 34) has noted that this view of the socialization process essentially is "mirror-like"; that is, children are believed to simply reflect the culture known and used by the adults of their group. Allport points out that this assumption ignores the many ways infants and children actually are known to take a creative part in shaping the process of learning culture.

Among others, Deutsch (1963), Frank (1938), Martin (1960), Maslow (1954), and Mead (1949b) have offered examples of the ways children take a creative part in the socialization process. In developing these ideas, Goodman (1964, 1967) proposed that children can be said to *generate* attitudes and values out of the cultural information made available to them by adults. Bruner's studies (1956, 1966) of cognitive growth also have provided empirical evidences that children can and do take an active role in the process of learning culture.

Thus the generative function of the socialization process may be defined as comprising all of the ways children reflect upon, think about, and sort out the content of culture available to them, in order to develop for themselves a cognitive map of adult culture.

(E) *Anticipatory Functioning.* Each one of the 128 cultures in the sample provides specific preliminary clues and indications of the ways children will be expected to act as adults.[19] An example would be found

[19] Early drafts of this work used the concept of *prefigurative* functioning as first defined by Margaret Mead in a discussion of the cultural determinants of sexual behavior (Mead, 1961). Mead has subsequently developed a theory in which she employs the concept of *prefiguration* in conjunction with the concepts of *cofigura-*

in the enculturation process among the eastern highland peoples of New Guinea (Berndt, 1962, p. 92). Here adult behavior has elements of what Americans would term "callousness" and "hostility" toward others, including infants and children. The children most admired by adults are those frequently commanding wide attention with violent temper tantrums, through their bullying of younger and weaker playmates, and those regularly given to excessive displays of aggressive posturing and swaggering. Children regularly carrying tales to adults also are widely admired, since this activity is viewed by adults as a normal and desirable feature of adult life. Children behaving in these ways are believed to have a future as successful adults, particularly in their potential as fighters and mothers. Thus such behavior by children is analogous with specific adult acts. Although these acts by children are not adult in their form, the significant attributes of such acts anticipate approved and expected adult behavior. In acting in these ways children have an opportunity to rehearse the behavior expected of admired and respected adults.

(F) *Developmental Functioning.* After birth human growth and maturation processes are related directly to the socialization process. There is a general human biological impetus toward "completion" of the basic direction imparted to the organism through hereditary materials. The pioneering studies of Piaget (1929, 1947, 1954, 1955; Piaget and Inhelder, 1948) and Gesell and Ilg (1950), among others, have made possible the understanding that cultural transmission generally occurs in a manner calibrated with the growth and body development of infants and children; that is, as Mead (1947) notes, each major act of cultural learning occurs in a definite relationship to a child's degree of biological readiness for the act. Further, the specific relationships between acts of cultural learning and children's biological readiness to learn create special situations in which deutero functioning of the socialization process can and often does occur. In other words, while most children learn to walk, and so on, the way they learn and the times they are permitted, or forced, to learn motor skills in reference to their actual biological capacity for walking and so on may become a vital fact in their participation in the cultural transmission process.

---

tive and *postfigurative* styles of cultural learning. In her comment on this discussion, Mead suggested that the concept of *anticipatory* functioning be substituted for *prefigurative* functioning. I believe it vital to maintain the integrity of the theoretical paradigm Mead has developed over the past fifteen years (Mead, 1956b, 1957, 1959, 1965, 1970) concerning the ways the learning of children reinforms adults' understanding of their culture. Hence, I have substituted the term *anticipatory* for *prefigurative* in developing a conceptual scheme for the socialization process.

Thus Balinese babies regularly are encouraged by adults to walk before crawling, so that infants suffer frequent losses of balance. This "reversal" of a human developmental process (crawl, then stand assisted, then stand unassisted, then walk alone) appears (Mead, 1947) to be related directly to the adult Balinese preoccupations with personal disorientations of any type, including those caused by use of alcohol and travel in modern vehicles. Mead suggests that the adult Balinese preoccupation with personal disorientation is made into a Balinese child's preoccupation through a cultural structuring of developmental skills in such a way that infants become concerned, from their being forced by adults to act in ways they cannot, with an acute sensing of imbalance and motion disorientation.

(G) *Linguistic Functioning.* *Homo sapiens* infants usually are born with a capacity for learning and using language. Hence, it is "biologically normal" for humans to learn and to use language, provided they have regular access to linguistically competent speakers of a language. Children will learn and use the language, or languages, spoken by the adult members of the culture into which they are born, provided adults expect them to do so. It is "normal" for children to undergo a series of complex transitions in the process of language learning and use prior to becoming linguistically competent.

In each culture infants and children must learn to shape correctly the minimally distinctive units of sound (or phonemes) used by competent speakers of the languages being learned. Too, children must learn an intricate system for joining phonemes into words (or morphemes). Children also must learn a grammar that controls the specific ways adult speakers of a language join morphemes to make cognitively useful utterances. Thus language learning does not simply involve acquisition, in a stimulus-response fashion, of a large collection of sounds, made into patterns peculiar to a language. Human languages involve exceedingly complex sets of rules that determine for speakers the specific sequences and patterns of speech sounds to be employed in particular social situations. Such rules must be mastered and used by each child, essentially on their own recognizance, that is, without adult instruction, direction, or teaching.

In the first stages of language learning, lasting from birth through about nine to eleven months of age, early speech sound patterns seem primarily to be functions of body maturation, accompanying specific developmental changes in the infant's anatomical and central nervous systems. The phonemic productions of infants are significantly influenced by the positions and postures used by a culture in the typical care of younger children. The first sound productions seem not to be related to the phonemes typically used in the languages of adults caring for infants. Sometime after nine to eleven months of age, infants begin to imitate and repeat the phonemic productions that emerge in their own spontaneous play. At about this same time human infants begin to make some recurring utterances of simple phoneme combinations that can be interpreted by adult speakers as having cultural and social meanings.

The first morphemes used by infants tend to be "total utterances," that is, a way of communication that has a broad sweep of complex cultural meanings. The growth of total utterance morphemes used meaningfully accelerates after the age of twelve to fifteen months.[20] At about eighteen to twenty months of age infants begin to put morphemes together to form primitive two-word sentences. The number of two-word sentences used increases markedly between eighteen to twenty and twenty-four to thirty months. During this time (that is, eighteen to thirty months) children combine and use two-word sentences in a fashion unique to them and not at all typical of the adult languages spoken in the group into which they have been born. In other words, eighteen- to twenty-six-month-old children use a grammatical system quite distinctly different from any particular adult grammatical system.

At about twenty-four to twenty-six months of age children begin to use a few hierarchical sentences, that is, two-morpheme sentences to which other morphemes have been added. By thirty months of age children are making increasingly greater use of hierarchical sentences, particularly those involving only three morphemes. At thirty-six months children have begun to more regularly employ four-morpheme hierarchical sentences, and at forty-eight months are using some six- to eight-morpheme sentences. At five years most children regularly are using multiple-morpheme sentences including some with combinations of ten and more morphemes.

During the time from the first use of hierarchical sentences (twenty-four to twenty-six months) to about forty-eight months of age most children "overregulate" morphemes as they use the child grammar system in handling multiple-morpheme sound productions. This often occurs through use of inappropriate plurals ("dogses," "foots," "manses," and so on) and substitutions of "strong verbs" inflected for their past tenses.

Studies of the linguistic overregularization process in two- and three-year-old children tend to agree that these language uses represent a consistent effort by young children to search out consistent patterns in language, to allow them access to new and more complex ways of speaking. When children reach the point of consistently using four- and five-mor-

[20] For discussions of and citations to research on language learning see Carroll (1964), Greenberg (1963), Hymes (1964), Bernstein (1961), Bright (1963), Jakobson and Halle (1966), Lenneberg (1964, 1967), and McNeill (1970a). See also the discussion and references in Chapter 10, Williams (1972).

pheme sentences, they also begin to adopt the grammar forms of the languages spoken by the adults caring for them. The details of this transition are not well known at present, but the available evidence seems to indicate the transition involves at least a half dozen or more stages.

It is through the linguistic functioning of the socialization process that infants and children come to have access to the "cognitive maps" of the adults who care for them, that is, to know what others know and expect them to know. It is through language that infants and children can be reached by and respond to those individuals, social institutions, and groups involved in the socialization process. It is through language that children learn of the conceptual scheme for socialization known and used by adults caring for them. It is through language that children come to develop the uniquely human psychological forms and processes of perception, attention, interest, memory, dreams, imagination, and symbolic representations, to deal with contrasts between the actual (real) and the expected (ideal), the tangible and intangible, and to differentiate between reality and fantasy. And it is through the linguistic functioning of the socialization process that children are able to take on the cultural preoccupation of adults, to be directly and actively instructed by personal agents of socialization, to acquire, through implicit, deutero and anticipatory, and other functions of the socialization process, the standard cultural meanings and meaning variations that comprise adult cultural preoccupations. It should be noted that major aspects of the human psychological attributes noted above (perception, imagination, and so on) also are gained in a paralinguistic manner (cf., Hall, 1964) as well as in kinesic (Birdwhistell, 1970) and other nonverbal forms (cf., Williams, 1966).

DISCUSSION: SOCIALIZATION PROCESS CONCEPTUAL SCHEME

In summary, I have proposed that a conceptual scheme for the socialization process may be conceived as consisting of seven structural features (*personal agents, impersonal agents, formal social groups, informal social groups, a socialization conceptual scheme, total human biological equipment, internal conditions of culture*) and seven functional features (*explicit functioning, implicit functioning, deutero functioning, generative functioning, anticipatory functioning, developmental functioning, linguistic functioning*).

There are, of course, a variety of ways to conceive of the socialization process. I have chosen to use a structural-functional model in construction of a socialization conceptual scheme. There may be errors in the scheme outlined briefly here, arising from the limited sample of enculturation processes used in the development of the scheme, from failure to differentiate clearly between the proposed structural or functional features, or from the logical problems in use of the structural-functional model for analysis of social and cultural behavior. Too, there may be more (or less) structural and functional features than are presented in this conceptual scheme.

It seems clear that an understanding of the origins, development, and nature of the socialization process cannot proceed without development of some type of conceptual scheme. In the absence of such a scheme, or schemes, we shall continue to employ concepts with variable meanings and to employ research procedures and methods appropriate to studies of some kinds of human phenomena, but not to the study of the socialization process.

It may be helpful to offer a socialization process conceptual scheme mnemonic device for use in further discussions. If each one of the structural parts of the conceptual scheme outlined above is diagrammatically represented as a closed square, the parts then can be allied in the form shown in Figure 1. Then it is possible to represent each of the functions of the conceptual scheme as a boundary entity forming and maintaining the structures in Figure 1. In this fashion the functions of the socialization process conceptual scheme are represented as binding the several structures, one to the other, as the consequence of the operation of the socialization process through time. The incompletely formed squares on the four sides of this conceptual scheme, and the one missing function, are meant to suggest that there may well be unrecognized structures and functions of the socialization process. Figure 2 offers two alternate mnemonic devices, utilizing conceptual scheme arrangements in which the structural features are noted as circles and the functional features are noted as either the points of intersections of structural features or as the several axes about which structural features are arranged. One theoretical advantage of the use of a square representation of a socialization process conceptual scheme is that it allows for greater freedom in seeking added structures and functions.[21]

A problem that can arise in construction and use of a conceptual scheme mnemonic device is that precision can be implied in graphic representations. The structural and functional features proposed here as comprising a socialization process conceptual scheme have been abstracted from a limited sample of enculturation processes. These features have not been subjected to any form of measurement, that is, they have not been "weighed," "tested for hardness," or measured for "fragility," "density,"

[21] For instance, Gerald Erchak has suggested, in a comment on a draft of this work, that an eighth structural feature in Figure 1 be designated as "human ecology."

Figure 1. A Mnemonic Device for a Socialization Process Conceptual Scheme.

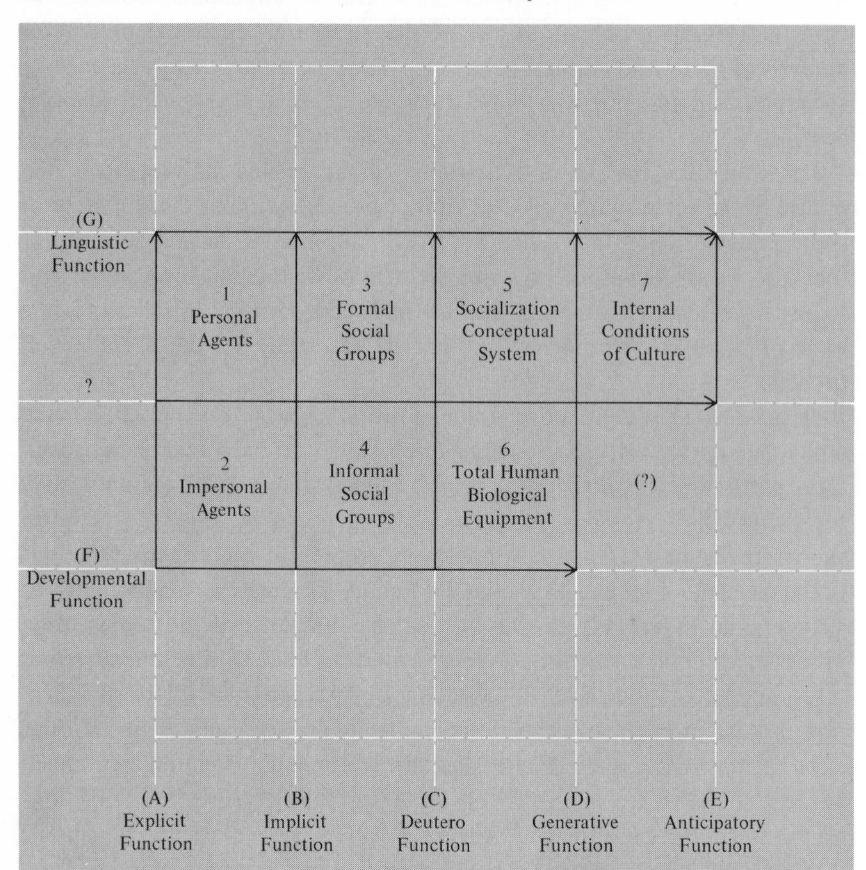

"boiling point" and so on. Even if sufficient data were available, genuine mathematical procedures for making precise measurements of socialization process structural and functional features remain to be developed. When such procedures are developed, they may significantly alter any existing conceptual scheme. However, as Wiener (1950, p. 26) has noted, until such a time it is both more honest and scientifically accurate to give a descriptive account of the gross appearance of a phenomenon and to carefully avoid leaving any impression of use of analytic precision and control.

Finally, the development of a socialization process conceptual scheme can lead to more efficient field studies of enculturation processes, since such a scheme not only calls attention to many features of cultural transmission that might be overlooked, but also will allow a particular enculturation process to be viewed as part of a human socialization process. Lacking a conceptual scheme base, existing guides for field study (cf.,

Figure 2. Two Alternative Socialization Process Conceptual Scheme Mnemonic Devices.

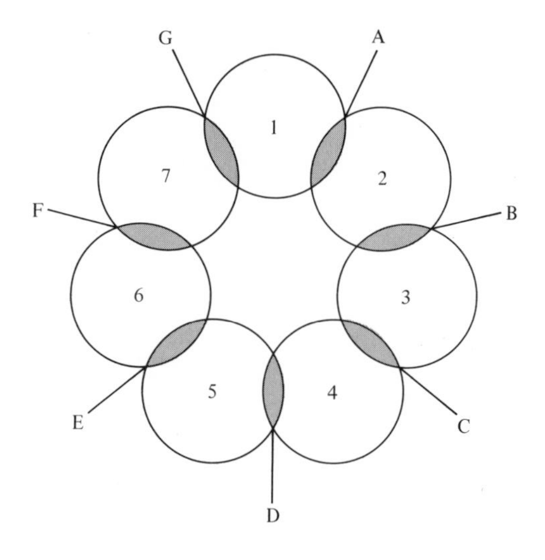

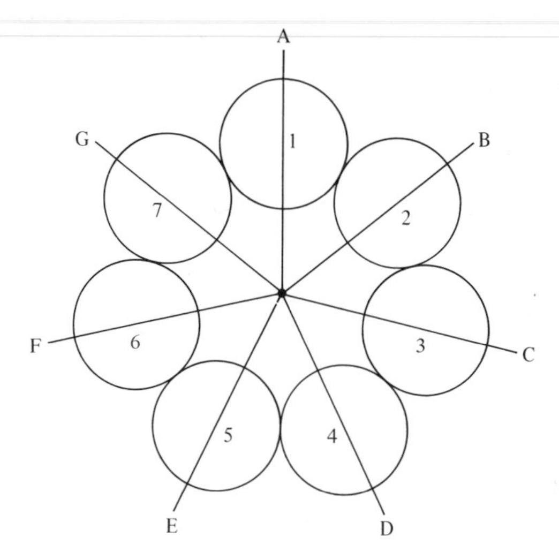

Murdock *et al.,* 1961; Hilger, 1960; Seligman, 1951) cannot lead observers to be alert to important differences in and between enculturation processes that may lead to new theoretical insights.

For instance, in a study of the Papago Indian enculturation process (Williams, 1958), I had no conceptual scheme for the socialization process available for use in the field. Some years later, after developing a preliminary version of a socialization conceptual scheme, and while in the course of study of the enculturation process of the Dusun of northern

Figure 3. A Comparison of Papago and Dusun Enculturation Processes.

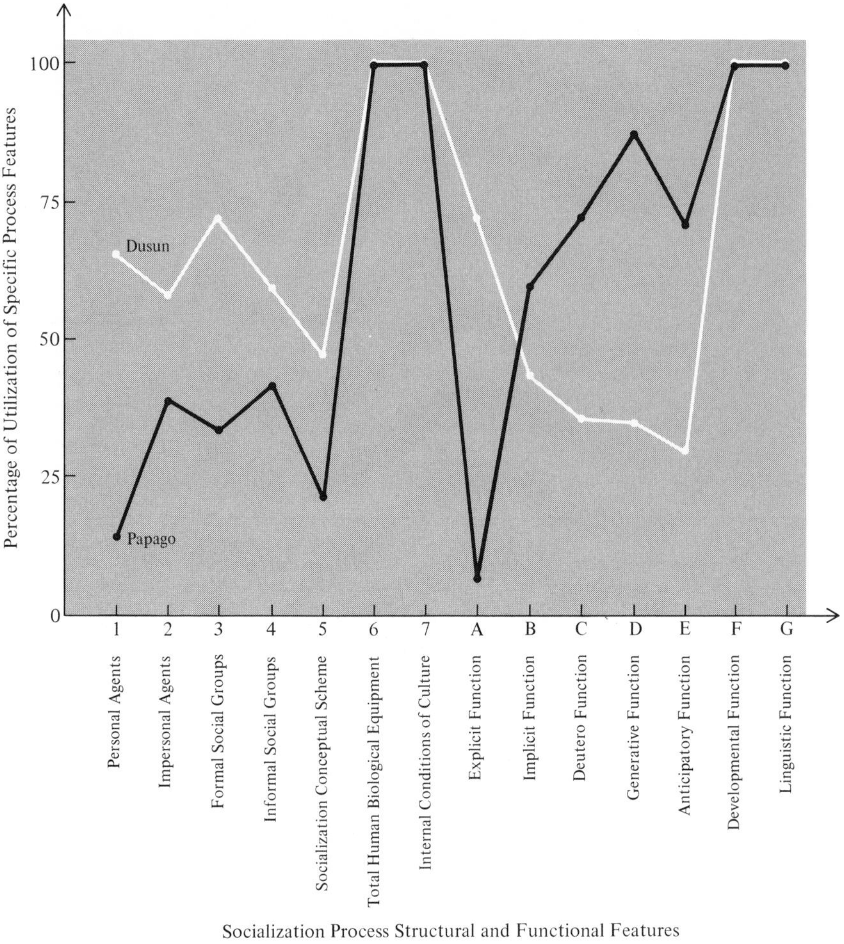

Percentage of Utilization of Specific Process Features

100
75
50
25
0

Dusun
Papago

1 Personal Agents
2 Impersonal Agents
3 Formal Social Groups
4 Informal Social Groups
5 Socialization Conceptual Scheme
6 Total Human Biological Equipment
7 Internal Conditions of Culture
A Explicit Function
B Implicit Function
C Deutero Function
D Generative Function
E Anticipatory Function
F Developmental Function
G Linguistic Function

Socialization Process Structural and Functional Features

Borneo (Williams, 1969), I realized that the Papago cultural transmission process was quite different from many other such processes in a number of significant respects. Papago adults rarely can be seen "doing," or directly telling children anything concerning behavior or culture. One waits for long periods before anything occurs that could be said to resemble cultural transmission. Yet Papago children regularly learn their culture. The usual models of human learning are quite difficult to apply to Papago enculturation (see Williams, 1972, Chapter 3), not because the Papago are unique among humans, but because such models do not account for the alternative ways humans may learn their cultures.

In terms of the conceptual scheme developed above, the Papago enculturation process operates primarily through children generating, from implicit, deutero, and anticipatory types of functions, the ways they are expected to behave as children and adults.

In contrast, the Dusun enculturation process operates through personal agents providing a great deal of explicit instruction, dependence upon use of impersonal agents of enculturation, formal and informal social groups, and a generally known and fairly explicit adult understanding of the local version of a socialization process conceptual scheme. Dusun adults place much less dependence on implicit, deutero, and anticipatory functions than do Papago adults, choosing to rely more upon their own understandings of and abilities to transmit Dusun culture than to allow their children to generate from their enculturation experiences a conception of Dusun culture.

These general contrasts between the Papago and Dusun enculturation processes are noted schematically in Figure 3. This figure is intended only to provide a descriptive comparison of these processes and is not intended to portray any precision of measurement of the structural and functional features involved.

ORIGIN AND DEVELOPMENT OF THE SOCIALIZATION PROCESS

Among others, Washburn (1950, 1958, 1963), Goodall (1964), and Oakley (1968, 1969, 1970) have clearly demonstrated it is not useful to continue searching for critical points of in toto form transition in primate evolution, or in seeking crucial events, such as toolmaking, to differentiate *Homo sapiens* from nonhuman primates.

Too, Washburn's hypothesis (1958, 1963) that behavioral changes may precede major anatomical changes in evolution greatly expands the perspective it is possible to adopt when considering hominid evolution, for it is no longer necessary to demonstrate that a kind of whole anatomical form exists before postulating the existence of a type of behavior. Thus it is possible now to conceive of a *Homo sapiens* behavioral process, such as socialization, as a property of behavior in primate forms that did not resemble *Homo sapiens* in structure.

Those persons concerned with study of the consequences of the socialization process often seem to adopt a "crucial point hypothesis" when they assume in their research that this process is solely confined to *Homo sapiens.* Similarly, the attitude that it is not useful to search for the origins of the socialization process in the behavior of other animal forms and especially nonhuman primates, because other animals cannot be shown to possess "culture," also reflects a crucial point bias. In any instance such biases seriously limit understanding of the nature of the socialization process.

The procedures developed by Hockett (1958, pp. 569–586) and Hockett and Ascher (1964) in discussing the origin and development of language are useful in considering the transtemporal dimension of the process of socialization. Thus when the preliminary conceptual scheme outlined above is applied to data of contemporary animal behavior, it becomes obvious that some features of the *Homo sapiens* socialization process also are formal (that is, regular) properties of behavior among some contemporary animal forms. For instance, the behavioral repertoire of other *Eutherians* (placental mammals) clearly includes features of the socialization process of *Homo sapiens*. Thus Darling (1937) has documented that a complex, direct, intergenerational information exchange regularly takes place among red deer in a nonsymbolic, inarticulate form, which occurs when adult female deer repeatedly lead young animals through the experiences of a variety of patterned behavior forms that are specific to a given environment and characteristic of that animal species. Suggestions also have been made recently by a number of ethologists (cf., Eibl-Eibesfeldt, 1970) that this kind of direct intergenerational, inarticulate information exchange occurs among a wide variety of vertebrates, including fish, amphibians, reptiles, and birds.

This does not mean and should not be taken to imply that some contemporary placental mammals or other vertebrates have or transmit culture. It means only that some specific behavioral features that seem to comprise the *Homo sapiens* socialization process possibly may be discerned in other life forms, some closely, others more distantly related to *Homo sapiens*. Among life forms other than the members of the primate order the task of tracing out precursors of the socialization process becomes complicated by questions of empirical evidence used, the style of logic, and the methodology of study, and as such, properly is the subject for a separate discussion.

However, if a search for the origin and development of the socialization process is narrowed only to *genera* of the primate order, it becomes more possible to conceptualize the broad outlines of the genesis of that process.

There is a general agreement now among students of primate evolution that this order evolved from *Insectivora* forms in the early Paleocene epoch, or between 60 and 70 million years before the present. There also is broad agreement that in the succeeding millions of years, a number of major biological adaptations and radiations occurred among the evolving primates. Two decades ago Washburn (1950) identified five such evolutionary adaptation-radiations and associated each one with a general evolu-

Figure 4. Some Major Primate Evolutionary Adaptations and Radiations.

Geological Epoch (millions of years)	Beginning Before Present	Extent	Primate Forms	Major Adaptations and Radiations		Ecological Relations
Pleistocene	3	3	Homo sapiens	5th Culture	terrestrial life; change in food sources; change in locomotor pattern; a change in social behavior patterns; development of system of learned, shared, patterned, and transmitted behavior manifest in act and artifact	Cultural Ecology
Pliocene	10	7	Homo	4th Bipedalism	terrestrial life; exploitation of new food sources; a continuing change in the locomotion pattern; a change in the social behavior patterns; a continuing change in senses	Transition from a Natural to a Cultural Ecology
Miocene	25	15	Hominidae		changes in dentition; a change in locomotor pattern; an increased range of locomotion; a change in the social behavior patterns	
Oligocene	36	11	Hominoidea (Apes)	3rd Brachiation	a continuing change in the senses; change in the locomotion pattern; an increased range of locomotion	Natural Ecology
Eocene	55	19	Cercopithecoidea (Monkeys)	2nd Special Sensory Reorganization	stereoscopic vision; middle ear changes; a change in the size and structure of the brain; a reshaping of the skull; an increased range of locomotion	
Paleocene	70	15	Prosimii (Lemur-Tarsier) (Insectivores)	1st Grasping Adaptation	aboreal life; exploitation of new food sources; changes in dentition; changes in the middle ear	

Source: After Washburn, 1950; Le Gros Clark, 1962; Campbell, 1966.

Note: This figure is meant only to be suggestive of broad changes and time levels, and is to be read as an extremely simplified version of a very complex series of biological and behavioral interrelationships in the course of evolutionary time.

tionary time level and structural type. Washburn's conceptual scheme subsequently has been modified in the works of LeGros Clark (1962), Campbell (1966), and others. Figure 4 summarizes Washburn's modified scheme. The time levels and structural types in this figure are based on the several available estimates for each of the major primate adaptations-radiations. This figure does not represent in any way a definitive portrayal of primate evolution. It is intended only as an *aide memoir* for readers of this discussion. However, such a summary can provide a point of departure for discussion of the origin and development of the socialization process:

1. There was a period when evolving primate populations lived regularly in and under a natural ecology, that is, when the system of mutual relations between evolving primates and their environments were not shaped, altered, or directed by any synthetic, processed, or acquired means.
2. There was in primate evolution a transitional period from a natural to a cultural ecology, that is, a time when the evolving primate populations began, however fitfully and incompletely, to synthetically shape and direct the relations between themselves and their environment.
3. There was in primate evolution a time when a cultural, as opposed to a natural, ecology generally obtained for some primate populations.

There are continuing disagreements among research specialists concerning the specific times that might be associated with each of these postulates. However, for the sake of this discussion, it will be assumed that the transitional period from a natural to a cultural ecology began at about the time of the transition from *Hominoids* to *Hominids,* that is, at approximately 15 million years before the present, and lasted until the beginning of the transition from the *Hominids* to *Homo,* at approximately 2 million years before the present. This provides for at least a 13-million-year period within which it is possible to initiate a search for precursors of the socialization process.

It may be reasoned also that it was in this transitional period from a natural to a cultural ecology that an evolving biology began regularly to interact with culture. Further, it may be speculated that the interaction between evolving hominid biology and culture could have activated a powerful mutual feedback process. Thus a purely biological evolutionary process began to be altered because of cultural changes, while in turn, cultural changes were affected by changes taking place in hominid biology. Time intervals between major changes in hominid biology and culture may have contracted sharply in the operation of a mutual feedback process. It also may be reasoned that in time, the feedback process came to have an existence of its own and to find a locus in the process of socialization.

Such a conjectural model can be misunderstood without careful exposition of the probabilities involved in its occurrence and at least a summary accounting of the likely course of events that have been proposed. It sometimes has been argued that living matter is so complexly organized that it is improbable that it could have developed on earth. If the process of *biopoiesis* on earth were specified as so improbable that it could happen only once in 1,000 times, the probability of its not happening is 999/1,000. Yet if 1,000 tries were to be made at such an event, the probability of the event not occurring drops to $(999/1{,}000)^{1{,}000}$ or 0.37. The probability of such an event taking place in 10,000 tries practically is inevitable. In the interactions between evolving hominid biology and culture, given (1) the length of time involved, (2) the particular nature of evolutionary hominid biology and experiences, and (3) the regular occurrence of non-random events (or tries) toward a mutual feedback process, it would seem highly probable that this process gradually would come to have a distinct existence and to find a specific locus in primate life.

The course of such a transition may be outlined briefly through use of a partly symbolic paradigm:

1. A hominid population or populations, *W*, living in a natural ecology, *NE*, could survive more readily in a cultural ecology, *CE*.
2. *W* did not regularly produce or employ any features of a *CE* while living under conditions of *NE*.
3. A *CE* is produced by an interaction synthesis process between hominid biology, *A*, and hominid experience, *B*.
4. A hominid population, or populations, *X*, evolve from W^N with genetic frequency changes giving *X* the capacity to synthesize a *CE* from the mutual interactions between *A* and *B*.
5. *X* produces some features of a *CE* from an *A* and *B* mutual feedback process.
6. As the *X* population or populations produce a *CE* from *A* and *B*, it gains a naturally selective evolutionary advantage over population W^N and so replaces it in a habitat.
7. A hominid population or populations, *Y*, evolve from X^N, with additional genetic frequency changes giving *Y* increased capacity to synthesize features of a *CE* from the mutual feedback interaction between *A* and *B*.
8. *Y* produces a *CE* from an *A* and *B* mutual feedback process.
9. As the *Y* population or populations produce a *CE* from *A* and *B*, it has a naturally selective evolutionary advantage over population X^N and so replaces it in a habitat.
10. *Y* transmits some of the features of *CE* it synthesizes from *A* and *B* to its young Y^1, who acquire some *CE* without their own synthesis from *A* and *B*; Y^1 also matures with and uses the capacity to synthesize *CE* from *A* and *B*.
11. Y^1 produces added *CE* features from an *A* and *B* mutual feedback process.

12. Y^1 transmits some of the *CE* it synthesizes from *A* and *B* to its young, Y^2. Y^1 also transmits some of the *CE* it has acquired from *Y* to Y^2. Y^2 thus possesses features of a *CE* that are cumulative in nature.
13. Y^2 produces added features of a *CE* from an *A* and *B* mutual feedback process.
14. Y^2 transmits some of the *CE* it synthesizes from *A* and *B* to its young, Y^3. Y^2 also transmits some of the *CE* it has acquired from *Y* through Y^1.[22]
15. A hominid population or populations, *Z*, evolve from Y^A, with additional genetic frequency changes giving *Z* increased capacity to synthesize a *CE* from the mutual feedback interactions between *A* and *B*. *Z* also possesses knowledge of features of a *CE* synthesized by its predecessors and culturally transmitted to it.
16. *Z* produces a *CE* (1) from an *A* and *B* mutual feedback process and (2) through integrating acquired features of a *CE* with the products of $A + B$ synthesis. The amount and varieties of *Z*'s *CE* differ significantly from that produced by any previous hominid populations.
17. As the *Z* population or populations produce *CE* from *A* and *B*, it has a naturally selective evolutionary advantage over population Y^N and so replaces it in a habitat.
18. *Z* transmits some of the *CE* it synthesizes from *A* and *B* to its young Z^1 who acquire some *CE* without synthesis from *A* and *B;* the amount of *CE* transmitted by *Z* to Z^1 and the amount of *CE* acquired by Z^1 are significantly increased over the amounts of *CE* previously transmitted and acquired by any hominid population. Z^1 also matures with and uses the capacity to synthesize from *A* and *B*.
19. Z^1 lives more in and under conditions of a *CE* than any previous hominid populations, that is, W^N, X^N, Y^N, and *Z*.
20. A hominid population, or populations, Z^N, evolve from Z^1 with additional genetic frequency changes giving Z^N increased capacity to synthesize *C* from the mutual feedback interaction between *A* and *B*. Z^N produces and transmits both the *CE* it has synthesized and acquired from preceding generations and lives more in and under conditions of *CE* than any previous hominid populations.

It is important that this paradigm be understood only to be illustrative of a *conjectural model* for the origin and development of the socialization process. It is not proposed or in any way meant to be used as a basis for calculations of gene frequency changes among evolving primates, for noting

[22] Mead has noted (personal communication, May 31, 1971) that this conjectural model does not allow for the effects that the young may have had upon adults in the generational process of cultural transmission.

specific time sequences in biological or cultural evolution, nor is it to be taken as a specific explanation for the origin of particular cultural features such as social organization, language, and so on. The primary purpose of such a symbolic paradigm is to draw attention to the ways an interaction synthesis process between evolving hominid biology and learned, shared, patterned, and generationally transmitted experiences may have developed into a cultural ecology.

The vital question in such a model is whether, in fact, hominid biology and experience really could interact mutually and in such a manner as not only to produce changes in each of the components of biology and culture but also to generate from these interactions a mutual feedback process that came to have a separate existence. There now seems to be little disagreement that there are specific effects on *Homo sapiens* biology due to cultural practices and vice versa. Such interrelations are well documented (cf., Chapple, 1970). It seems reasonable to assume that these reciprocal effects may have been operative for a very long period.

Until recently, culture theorists have been divided into two major schools. On the one hand, there has been the theoretical position, derived from the eighteenth- and nineteenth-century traditions of naturalism and positivism, which depicts culture as an autonomous, superpsychical, and superorganic entity, subject only to its own laws, stages of development, and internal dynamics. On the other hand, there is the theoretical view that has its roots in the humanistic traditions of the European Renaissance and the rationalism of the eighteenth-century philosophers of the Enlightenment, which portrays culture entirely as a product of human discovery and creativity and therefore as fully subject to human control and direction. However, neither of these theoretical positions has been very helpful in considering the ways human culture and biology evolved to their present forms and interrelationships. The superorganic position faces a nearly insuperable barrier when it comes to the questions of the specific origins of an autonomous system of culture; one must either invoke a supernatural intervention in primate evolution or use a variety of teleological arguments, which essentially imply that culture is part of a grand design, or plan, for human life presently unknown (and probably unknowable) to man. The humanistic theoretical position also has to face similar logical barriers to its use, for under a crucial point hypothesis, it has been exceedingly difficult to conceive of the ways hominid forms so very different in their whole body appearances from *Homo sapiens* could possibly have discovered or created anything so complex as the culture, or even most features of the culture, of *Homo sapiens*.

Recent trends in cultural theory provide indications that these two theoretical positions may be combined broadly into a form that notes the roles of intelligence, cognitive freedom, and the acts of individuals in forming and determining culture, while also acknowledging the internal dynamics and logic of a system of evolving intergenerationally transmitted knowledge comprised of an accumulation of individual acts and contributions (cf., Kroeber and Kluckhohn, 1952). In following such a modern synthesis of cultural theory, one quite important question is: In what manner and with what "mechanisms" did a mutual feedback process between biology and experience first appear, then undergo development and growth?

It may be proposed that there were at least three vital "mechanisms" involved in the appearance and development of a hominid biology-behavior mutual feedback process. These may be designated as (1) *a reflexive capacity,* (2) *a symbolic communication capacity,* and (3) *a capacity for learning through operant conditioning.*

The capacity for reflexion involves cognitive anticipation of future events on the basis of past and present experiences and acting in ways that will directly affect future events. A symbolic communication capacity involves productions, through behavior, or use of arbitrary sounds, of signs that suggest the products of reflexion and particularly reflexion concerning quantities, qualities, spatial positions, and relationships of past, present, and future events. A capacity for learning through operant conditioning involves an animal self-directing, or "rewarding," or "punishing," itself for its behavior. In operant learning, as distinguished from other kinds of learning (for example, stimulus-response, or "classical" learning), animals essentially teach themselves directly from their experiences, as well as being led through a learning activity by peers or adult animals.

The development of an interaction between hominid biology and experience perhaps would have depended upon these three mechanisms in this manner: Primates, increasingly more able to reflect upon the future consequences of their present and past behavior, would have come to have naturally selective advantages over other populations competing for places in the same natural habitat. A growing capacity to psychologically "rehearse" the future may not have been overly useful to a population while it still was confined to single, uncommunicative primates, except insofar as it contributed to the physical survival of solitary individuals and thereby ensured a reproductive continuity for the reflexive capacity within a population. However, once the products of individual reflexivity could, in some manner, be gotten from the "inside" to the "outside," that is, transformed from solely mental images to externally visible signs made in behavior or sounds that broadly represented individual reflexive products, then an evolving primate population would be able to share quickly and widely among themselves the several versions of the consequences of present and past behavior for the future of not only solitary individuals but the entire population.

The rise of a symbolic communication capacity provided for a further and very significant naturally selective advantage for those evolving primate populations that possessed it, for it was these groups that gained the edge of a "collective wisdom"; that is, through a widespread sharing of individual reflexive products, they could have the advantage of selecting and accepting the portrayals of the most reflexively adept individuals over the products of the less reflexively able individuals. An evolving primate symbolic communication capacity became the way whole primate populations could gain access to the individual psychological resources that were the most genetically "progressive" or more "intelligent." This means of gaining a naturally selective evolutionary advantage meant that the very slow process of diffusion of behavioral change through sex cell distribution in reproductive acts, which ordinarily occurs over a span of very many filial generations, was accelerated significantly through a wide sharing in only several filial generations of the cognitive productions of a great many of the most reflexively efficient members of a population.

In terms of the origin and development of the socialization process, it may be said that at the point when evolving primate populations, possessed of increasingly more effective capacities for reflexivity and symbolic communication of the products of reflexion, began to share reflexive products with their young the transition commenced from a natural to a cultural ecology.

Since these two developing capacities (reflexion and symbolic communication) probably were then as now linked to physical maturation processes, it is likely that some specific efforts by adult primates would have had to be made to bring the infants and young children of each generation into a developing adult system of symbolically communicated and shared reflexive productions. In the absence of such a cumulative, intergenerational sharing process, each new generation would have had to depend upon the reflexive productions that were derived and shared within that one generation. It may be proposed that the appearance of a process of intergenerational sharing with infants and young of cumulative reflexive productions through symbolic communicative means marked the appearance of a socialization process.

Recent studies of operant conditioning have noted that brain cell nucleic acid changes are produced in the learning process. It has been suggested as the consequence of such research that nerve cell nucleic acids possess most information learned through the lifetime of an animal. This hypothesis is based on the fact that the growth of nerve cells and nerve cell connections appears to depend upon the long-term responses these materials make to repeated experiential stimulations. Since all nerve cell growth basically involves the synthesis of nucleic acids, different kinds of nerve cell stimulation should be reflected in differing patterns of nucleic acids.

To test this particular hypothesis, Hydén and Egyházi (1964) conducted a highly innovative operant-conditioning experiment in which rats were permitted to reach into a cylinder to obtain food. Each subject was allowed twenty-five food extractions to demonstrate a preference for "handedness" (or "pawness"). If a subject extracted food with a right paw twenty-three out of twenty-five times, it was labeled as "right handed" and then placed in a situation in which it could obtain food only through use of the left paw. During the subsequent four days of the experiment each normally right-pawed rat made between 400 and 500 attempts to extract food with the left paw. At the end of four days nerve cell tissues were extracted from the cerebral cortex areas that are known to control handedness in rats. These tissues then were analyzed for mRNA content; this nucleic acid has been demonstrated to be integral in the cell physiology process in which proteins are specified in tissue growth and repair. Hydén and Egyházi were aware that if the forced transfer to left-handed food extractions had produced any changes in the mRNA content of cortex tissue, it would appear in the right area of the cortex. Their analysis of mRNA content demonstrated a significantly higher level of mRNA in the right cortex of each of the operantly conditioned rats. To conclusively demonstrate that a significant increase in mRNA had taken place as a consequence of the operant-conditioning experiment, Hydén and Egyházi examined the cortical tissue mRNA content of a control group of right-handed rats from an identical genetic strain to the rat subjects used in the experiment. It was found that there were no significant differences in the mRNA content of the tissues in right or left cerebral cortexes of rats in the control group.

These results, and others of a similar nature from subsequent experiments in operant conditioning and nerve cell nucleic acid changes, provide some support for the hypothesis that operant learning can result in pattern changes of nucleic acids and proteins in the cells directly involved. Such changes appear to be time dependent in every instance, that is, directly related to the amount of operant conditioning involved.[23]

The concept of operant learning provides a basis for vital departure from previous, so-called classical models of learning, in which various "external" influences were said to automatically produce behavioral re-

[23] In a review of brain chemistry and behavior research Cooper *et al.* (1970) have noted that this area of research is in a state of constant change with often contradictory results, theories, and methods. They warn that the "buyer" of such information should be wary of extending the conclusions of any particular study.

sponses in an organism. Operant conditioning research makes it possible to shift to a learning theory in which organisms direct their own responses to many different stimuli, whether these arise in the "outside" or the "inside" environments. And recent studies in behavioral physiology and genetics, which have used operant-learning procedures, seem to indicate that there are significant central nervous system nucleic acid changes as a consequence of such self-directed learning.

If such research proves to be scientifically sound, then the process of operant learning may have been a vital mechanism in the appearance, development, and fixing of a mutual feedback process between evolving primate biology and behavior. It may be reasoned that evolving primates, increasingly using capacities for reflexion and symbolic communication, could have benefited from a learning process that regularly and directly links the components of biology and experience. In other words, it may be suggested that at some point, perhaps before or after the appearance of hominids, operant learning contributed to such significant nucleic acid and protein changes as to lead to a linking together of primate biology and experiences in a process in which experience led to changes in biology while changes in biology reciprocally led to life experience changes. If operant learning had been a vital process in primate evolution well before the appearance of hominids, and there is no reason not to believe this was the case, then the essential development of a mutual feedback process may have awaited only the development of the reflexive and symbolic communication capacities. There is no present way, on the basis of available empirical evidences, to assign any priority to one or the other of these three mechanisms (that is, reflexion, symbolic communication, and operant conditioning). It is useful, however, to conceive these three mechanisms, and any others that may have been involved, as operating in a synergistic form, that is, as working together in such a manner that the sum, or total, effect was greater than the sum of each of the three mechanisms working independently. It may be proposed that it was from this synergistic sum that the mutual feedback process between evolving primate biology and experiences derived its impetus to a separate and lasting existence. It may be proposed also that it was from such a synergistic relationship and sum as these continued over a very long period that a *Homo sapiens* socialization process could have developed.

SUMMARY AND CONCLUSION

This discussion has been concerned with a theoretical perspective for the socialization process. It began with a brief review of socialization definitions and by noting some problems in socialization research. Then a conceptual scheme was presented in an effort to provide some transcultural and holistic dimensions for understanding the socialization process. This discussion was followed by an account of a conjectural model for considering the transtemporal dimension of the socialization process.

The preliminary conceptual scheme and conjectural model developed in this discussion can be used to stimulate discussion and prompt further investigations and reviews of data, methods, and theory used in study of the socialization process. The scheme presented here is only one of a number of different ways the socialization process can be conceived. Given its present limitations, the usefulness of this scheme must be determined through its applications in the conduct of future research.

This conceptual scheme may prove to be of value in the course of field studies of the enculturation process, since to this point in time, such studies have been handicapped by a lack of a conceptual scheme for ordering data. A conceptual scheme for the socialization process may provide field workers with a wider scope of concern as they gather data that match or contradict theoretical categories, extend and refine definitions, and reorder the conceptual arrangements employed in construction of a scheme.

This conceptual scheme also may be of use in the conduct of studies of nonhuman primate learning and behavior. There has been a large number of reports of recent field studies of nonhuman primate learning and behavior (cf., Altmann, 1962, 1967; DeVore, 1965; DeVore and Lee, 1963; Jay, 1968; Schrier *et al.,* 1965; Southwick, 1963; Washburn *et al.,* 1965). Such observations tend to be ordered into the conceptual categories of food getting, reproductive cycles and actions (including frequencies and systems of mating), territoriality, aggression and dominance behavior, and intraspecies communication. It also may be helpful in such studies to employ conceptual categories derived from a *Homo sapiens* socialization process conceptual scheme. This would have the value of broadening the analytic repertoire of observers of nonhuman primate behavior while avoiding use of the concept of "socialization" simply to denote social behavior observed in the context of a nonhuman primate group (cf., Kummer, 1968, pp. 298–308).

Individuals especially concerned with field studies of nonhuman primate behavior seem to proceed in their studies with a basic assumption that because of their many morphological similarities to fossil forms, living primates can be assumed to be not too dissimilar to, and therefore to be representatives of, ancestral stages in human evolution (cf., Campbell, 1966, p. 59). Perhaps another way to state this assumption would be to note that observers of nonhuman primate behavior undertake such studies because they believe that when they record details, processes, and styles of behavior of different genera of living primates they also are recording features of learning and behavior as they possibly existed at different stages or periods in the process of human evolution (cf., Milner and Post, 1967).

A second assumption also seems involved in studies of nonhuman primate behavior. This assumption can be described best as one that treats nonhuman primate behavior as a less complex version of *Homo sapiens* behavior. For instance, it is not uncommon to find authors of nonhuman primate behavior studies regularly employing concepts such as "personality" (cf., Itani, 1959; Miyadi, 1964), "status," "role," and so on. To date, there has been no serious question concerning such uses in studies of nonhuman primates. The point is not whether nonhuman primates have personalities and so on, but whether when observers use such concepts, they are assuming that primates are behaving in essentially the same ways as *Homo sapiens* (and vice versa).

These two assumptions provide a means for students of nonhuman primate learning and behavior to conceptually "look both ways." On the one hand, using these assumptions, nonhuman primate behavior scholars assume that the animals they study represent living expressions of quite ancient phenomena, while, on the other hand, these scholars also can view their work as productive because it is concerned with the fundamentals of behavior so beclouded in *Homo sapiens* life by subsystems of culture such as technology, ideology, and language. In effect, students of nonhuman primate learning and behavior have the conceptual luxury afforded by simultaneous views to the ancestral primate past and a human present.

However, those persons concerned primarily with studies of contemporary human behavior must make the specific assumption that there are direct continuities between *Homo sapiens* and nonhuman primate behavior, that is, postulate that men today act regularly in ways that may be traced "backward" across the biological classification boundaries of the two superfamilies (that is, *Hominoidea* and *Cercopithecoidea*) of the primate suborder (*Anthropoidea*) in which it generally is considered human evolution occurred. There seems to be no question concerning the empirical evidences of organic relationships between *Homo sapiens* and living members of the *Hominoidea* superfamily (cf., Lancaster, 1967, p. 58). But there remains a major logical problem in use of this assumption by students of *Homo sapiens* behavior. This can be stated most clearly by noting that since some 25 to 35 million years of evolutionary development has occurred among the populations ancestral to the living members of the Cercopithecoidea and Hominoidea superfamilies, there is a distinct possibility that the forms of behavior, and particularly social life and psychological processes, of contemporary Old World monkeys and apes may well involve substantive differences of an order that highly qualifies any conclusions drawn on the basis of an assumption of direct continuity between such living primates and *Homo sapiens*.

This logical conflict does not seem to concern most students of nonhuman primate behavior, for under the two assumptions noted above, it is possible to see such continuities as existing and real. However, for persons primarily concerned with studies of *Homo sapiens* behavior, such continuities must be taken literally, since the analogies and similarities may be perceived only dimly from the perspective of the cultural and social anthropologist. A lively exchange between Washburn and Sahlins (cf., Washburn and DeVore, 1961; Sahlins, 1959, 1960) concerning whether *Homo sapiens* society developed after a subordination of sexual drives clearly reflects this problem of research. Galdston (1961, p. 289) aptly has noted and summarized this logical conflict in his comment concerning the differences in conceptual orientations of physical anthropologists and those concerned essentially with studies of *Homo sapiens* behavior: "The (physical) anthropologist seeks to reconstruct man's history from the 'before to the present'; the psychiatrist (cf., Freud, 1913) wants to understand the present in the perspective of the before." It does not seem reasonable at this juncture to expect individuals concerned essentially with studies of nonhuman primate behavior, or focusing primarily on studies of *Homo sapiens* behavior, to abandon the assumptions employed in their research. However, it does seem vital that these contrasting assumptions be accounted for in efforts to develop any valid transtemporal, transcultural, and holistic theory of the socialization process.

It may be said that these differing assumptions have been at the very center of the sharply critical responses of many scholars to recent popular accounts by Ardrey (1961, 1966, 1970) as well as to some studies by "professionally qualified" individuals (cf., Tiger, 1969). Ardrey uses all of the assumptions noted above and does not hesitate to expand them considerably in ways that seem generally unacceptable to many specialists. Thus Ardrey not only uses the assumptions that the behavior and learning of living nonhuman primates are simply less complex forms of *Homo sapiens* behavior and that *Homo sapiens* behaves today in ways that may be traced across the biological boundaries between hominids, Hominoidea, and Cercopithecoidea, he also extends these assumptions to include many examples of learning and behavior from varieties of animal forms (fish, birds, bees, and so on) distantly related to *Homo sapiens*.

The point that should be made here, in the context of Ardrey's work, is that he has evoked strongly critical response from many specialists through his proper identification but well-extended uses of their generally agreed upon research assumptions.

Those concerned with research on the transtemporal dimension of the socialization process must be prepared to face similar criticisms, since it is not really possible to speculate about or to discuss any type of conjectural model for describing the origin and development of the socialization process without running directly afoul of the assumptions currently used by specialists in studies of either *Homo sapiens* behavior or of nonhuman primate behavior and learning. For example, it may be predicted that few students of nonhuman primate behavior and evolution will find much to disagree with in the first two sections of this work. However, in the third part of the discussion, as a conjectural model is presented for discussion of the origin and development of the socialization process, it is to be expected that there may be a variety of objections, many of which will be derived from use of the unspoken assumptions employed by physical anthropologists.

It may be expected also that chief among responses to such a conjectural model will be the notation that there is really insufficient empirical evidence at hand to provide very much for meaningful comment regarding any of the "mechanisms," such as a reflexive capacity, a symbolic communication capacity, or a capacity for learning through operant conditioning, which could have been involved in the development of a mutual feedback process between hominid biology and life experiences. Similarly, it is to be expected that it may be said that such a conjectural model is merely a speculative exercise, since direct records of the origin of any behavioral process are not likely to be found in any paleontological or prehistoric records.

It may be argued, however, that a continuing failure to consider the transtemporal dimension of the socialization process, that is, questions concerning the origin and development of the socialization process, can only lead to further *ahistorical* theory development (cf., Clausen, 1968; Goslin, 1969; Shimahara, 1970) and therefore to still more inappropriate conclusions regarding the nature of the socialization process and its consequences in human life. Current socialization theory tends to be quite culture-bound and generally lacking in any time perspective. It would be most unfortunate if theory for research on the socialization process could not now flourish on the data and conclusions in works by students of nonhuman primate evolution and behavior. For instance, Simons and Pilbeam (*New York Times,* 1970) recently have suggested there are direct fossil evidences (rates of tooth wear and eruption) of significant differences in the maturation, and therefore in the learning processes, between the extinct ape *Dryopithecus indicus* and its hominid contemporary, *Ramapithecus punjabicus.* It should be possible for those concerned with the transtemporal nature of socialization to at the very least note that there seem to be no reasons now not to "look back" to at least a time some 12 to 15 million years before the present for the origin of that process, without becoming involved in an intricate and specialized argument on the merits and acceptability of fossil primate data. It would seem sufficient for purposes of initial formulation of theory to incorporate Simons and Pilbeam's conclusion that the differential rates of wear and eruption of teeth in fossil forms may well indicate existence of a prolongation of the time of learning in early hominids and therefore be a period in evolutionary time of which it could be said that there seems to be some evidence of the beginnings of some structural or functional features of a socialization process.

In conclusion, it would be of great assistance if students of nonhuman primate behavior and evolution, with their special knowledge and awareness of the assumptions used in their research, would develop or substantially contribute to a model for use in discussion of the origin and development of the socialization process. This kind of model development really ought not be left to colleagues essentially concerned with studies of *Homo sapiens* learning and behavior. Too, it may help substantially in development of a general theory concerning the socialization process if students of nonhuman primate behavior would employ some kind of conceptual scheme for socialization when they attend to what is believed to be the same process among gibbons, baboons, and so on. Providing the logical problems clearly are noted, there now seem to be no reasons for observers of nonhuman primate behavior not to employ categories derived from a socialization conceptual scheme as one means to order and to analyze data records. And, of course, students of the socialization process must now utilize some form of conceptual scheme in their field studies.

The articles on nonhuman primate learning and behavior included in this text employ a wide variety of ideas and terms to order data. In reading these accounts, if one uses the socialization conceptual scheme categories presented earlier in this discussion, some order and coherence tend to emerge; some authors are describing the ways particular kinds of nonhuman primate "informal social groups" and "personal agents" engage in the process of transmitting learned behavior. Some authors have remarked upon and compared the ways in which informal social groups are involved in learning of behavior in different types of nonhuman primates. There are repeated comments throughout these accounts that are directed to description and analysis of "developmental," "explicit," "implicit," "deutero," "generative," and "anticipatory" types of functions in the learning and behavior processes of a wide variety of nonhuman primates. But lacking a conceptual scheme for the socialization process, such comment tends to lose cogency and meaning through partly formed definition.

An application of a conceptual scheme for the socialization process, derived from data of *Homo sapiens* behavior, to studies of nonhuman primate behavior may prove fruitful for both kinds of research. But whatever the present results, it is clear now that any general theoretical perspective for the socialization process must incorporate not only transcultural and holistic data, but transtemporal data as well. Field and laboratory studies of nonhuman primate behavior promise to open new vistas in understanding the transtemporal dimension of the socialization process. There are many problems of research on the nature of the socialization process, but the ultimate rewards of final comprehension would seem to merit the effort.

Acknowledgments

I am indebted to a large number of persons whose publications I have used in developing my ideas concerning the nature of the socialization process. I have given specific credit to many of these persons elsewhere (Williams, 1969, p. 2). I wish particularly to thank Margaret Mead, Erika Bourguignon, Frank E. Poirier, Jr., and H. S. Morris for their helpful comments on this work.

References

AINSWORTH, M. D. S. (1967) *Infancy in Uganda.* Baltimore: The Johns Hopkins University Press.

ALLPORT, G. (1955) *Becoming.* New Haven: Yale University Press.

ALTMANN, S. A. (1962) "The Social Behavior of Anthropoid Primates: An Analysis of Some Recent Concepts." In E. L. Bliss (ed.), *The Roots of Behavior.* New York: Harper and Row, pp. 277–285.

———. (ed.) (1967) *Social Communication Among Primates.* Chicago: University of Chicago Press.

AMMAR, H. (1954) *Growing Up in an Egyptian Village.* London: Routledge and Kegan Paul.

ANDREW, R. (1962) "Evolution of Intelligence and Vocal Mimicking." *Science* 137: 585–589.

———. (1963) "The Origin and Evolution of the Calls and Facial Expressions of the Primates." *Behaviour* 20: 1–109.

ARDREY, R. (1961) *African Genesis.* New York: Atheneum.

———. (1966) *The Territorial Imperative.* New York: Atheneum.

———. (1970) *The Social Contract.* New York: Atheneum.

ASCH, S. E. (1955) "Opinions and Social Pressure." *Scientific American* 193: 31–35.

AUSUBEL, D. P. (1965) *Maori Youth: A Psychoethnological Study of Cultural*

Deprivation. New York: Holt, Rinehart and Winston.
BARNOUW, V. (1963) *Culture and Personality.* Homewood, Ill.: Dorsey.
BASEDOW, H. (1925) *The Australian Aboriginal.* Adelaide: Preece.
BATESON, G. (1942) "Social Planning and the Concept of Deutero-Learning." *Symposia, Conference on Science, Philosophy and Religion in Their Relation to the Democratic Way of Life* 2: 81–97 (reprinted in T. M. Newcomb and E. L. Hartly (eds.), *Readings in Social Psychology.* New York: Holt, 1947, pp. 121–128).
——— and M. MEAD. (1942) *Balinese Character: A Photographic Analysis.* New York: Special Publication of the New York Academy of Sciences, Number 2.
——— and J. RUESCH. (1951) *Communication: The Social Matrix of Psychiatry.* New York: Norton.
BERNDT, R. (1962) *Excess and Restraint.* Chicago: University of Chicago Press.
——— and C. BERNDT. (1964) *The World of the First Australians.* Chicago: University of Chicago Press.
BERNSTEIN, R. (1961) "Social Class and Linguistic Development." In A. H. Halsey, J. Floud, and A. Anderson (eds.), *Education, Economy and Sociology.* Glencoe, Ill.: Free Press, pp. 288–314.
BIRDWHISTELL, R. (1970) *Kinesics and Context: Essays on Body Motion Communication.* Philadelphia: University of Pennsylvania Press.
BOURGUIGNON, E. (1972) "Psychological Anthropology." In J. J. Honigmann (ed.), *Handbook of Social and Cultural Anthropology.* Chicago: Rand-McNally.
BRIGHT, W. O. (1963) "Language." In B. J. Siegel (ed.), *Biennial Review of Anthropology.* Palo Alto, Calif.: Stanford University Press.
BRIM, O. G., JR. (1968) "Socialization: Adult Socialization." In W. A. Wallis (ed.), *International Encyclopedia of the Social Sciences.* New York: Macmillan and the Free Press, vol. 14, pp. 555–562.
BRUNER, J. *et al.* (1956) *A Study of Thinking.* New York: Wiley.
———. (1966) *Studies in Cognitive Growth.* New York: Wiley.
BURTON, R. V. (1968) "Socialization: Psychological Aspects." In W. A. Wallis (ed.), *International Encyclopedia of the Social Sciences.* New York: Macmillan and the Free Press, vol. 14, pp. 534–545.
CAMPBELL, B. G. (1966) *Human Evolution.* Chicago: Aldine.
CARROLL, J. B. (1964) *Language and Thought.* Englewood Cliffs, N.J.: Prentice-Hall.
CHAPPLE, E. D. (1970) *Culture and Biological Man.* New York: Holt, Rinehart and Winston.
CHILD, I. L. (1954) "Socialization." In G. Lindzey (ed.), *Handbook of Social Psychology.* Boston: Addison-Wesley, vol. II, pp. 665–692.
CHOMSKY, N. (1968) *Language and Mind.* New York: Harcourt Brace Jovanovich.
CLAUSEN, J. (ed.) (1968) *Socialization and Society.* Boston: Little, Brown.
COHEN, Y. A. (1961) *Social Structure and Personality: A Casebook.* New York: Holt, Rinehart and Winston.
———. (1964) *The Transition from Childhood to Adolescence.* Chicago: Aldine.

———. (1968) "Macroethnology: Large Scale Comparative Studies." In J. Clifton (ed.), *Introduction to Cultural Anthropology.* New York: Houghton-Mifflin, pp. 402–488.

COOPER, J. R., R. H. ROTH, and F. E. BLOOM. (1970) *Biochemical Basis of Neuropharmacology.* New York: Oxford.

DARLING, F. (1937) *A Herd of Red Deer: A Study in Animal Behavior.* London: Oxford University Press.

DAVIDSON, D. (1926) "The Basis of Social Organization in Australia." *American Anthropologist* 28: 529–548.

DE BRUL, E. L. (1958) *Evolution of the Speech Apparatus.* Springfield, Ill.: Thomas.

DE LAGUNA, F. (1965) "Childhood Among the Yakutat Tlingit." In M. E. Spiro (ed.), *Context and Meaning in Cultural Anthropology.* New York: Harper & Row.

DENNIS, W. (1940) *The Hopi Child.* New York: Wiley.

———. (1941) "Infant Development Under Conditions of Restricted Practice and Minimum Social Stimulation." *Genetic Psychology Monographs* 23: 143–189.

DEUTSCH, M. (1963) "Nursery Education: The Influence of Social Programming on Early Development." *Journal of Nursery Education* 18: 191–197.

DEVORE, I. (ed.) (1965) *Primate Behavior: Field Studies of Monkeys and Apes.* New York: Holt, Rinehart and Winston.

——— and R. LEE. (1963) "Recent and Current Field Studies of Primates." *Folia Primatologica* 1: 66–72.

DOBZHANSKY, T. (1962) *Mankind Evolving.* New Haven: Yale University Press.

———. (1963) "Cultural Direction of Human Evolution: A Summation." *Human Biology* 35: 311–316.

——— and M. F. A. MONTAGU. (1947) "Natural Selection and the Mental Capacity of Mankind." *Science* 105: 587–590.

DRIVER, H. and R. CHANEY. (1968) "A Sixth Solution to the Galton Problem." *American Anthropological Association,* Bulletin 1: 35–36.

DUBOIS, C. (1956) "Attitudes Toward Food and Hunger in Alor." In D. G. Haring (ed.), *Personal Character and the Cultural Milieu.* Syracuse, N.Y.: Syracuse University Press, pp. 241–253.

EIBL-EIBESFELDT, I. (1970) *Ethology: The Biology of Behavior.* New York: Holt, Rinehart and Winston.

ELKIN, A. (1933) "Studies in Australian Totemism." *Oceania* 3 (Numbers 3, 4); 4 (Numbers 1, 2).

———. (1954) *The Australian Aborigines: How to Understand Them,* 2nd ed. Sydney: Angus and Robertson.

——— and C. BERNDT. (1950) *Art in Arnhem Land.* Chicago: University of Chicago Press.

ELKIN, F. (1960) *The Child and Society: The Process of Socialization.* New York: Random House.

EMBER, M. (1971) "An Empirical Test of Galton's Problem." *Ethnology* 10: 98–106.

ETKIN, W. (1963) "Social Behavioral Factors in the Emergence of Man." *Human Biology* 35: 299–310.

FIRTH, R. (1956) "Function." In W. L. Thomas (ed.), *Current Anthropology*. Chicago: University of Chicago Press, pp. 237–258.

FORTES, M. (1938) *Social and Psychological Aspects of Education in Taleland.* London: International African Institute.

FRANK, L. K. (1938) "Cultural Control and Physiological Autonomy." *American Journal of Orthopsychiatry* 8: 622–626.

FREUD, S. (1913) "Totem and Taboo." In *The Basic Writings of Sigmund Freud.* New York: Random House, 1938, pp. 807–930.

GALDSTON, I. (1961) "Comments." In S. L. Washburn (ed.), *Social Life of Early Man.* Chicago: Aldine, pp. 289–291.

GALTON, F. (1889) "Comment on a Paper by Edward B. Tylor: On a Method of Investigating the Development of Institutions: Applied to the Laws of Marriage and Descent." *Journal of the Royal Anthropological Institute* 18: 245–272.

GARDNER, P. (1966) "Symmetric Respect and Memorate Knowledge: The Structure and Ecology of Individualistic Culture." *Southwestern Journal of Anthropology* 22: 389–415.

GEERTZ, C. (1964) "The Transition to Humanity." In Sol Tax (ed.), *Horizons of Anthropology,* Chicago: Aldine, pp. 37–48.

GESELL, A. and F. ILG. (1950) *Child Development: An Introduction to the Study of Human Growth.* New York: Harper & Row.

GOODALL, J. V. L. (1964) "Tool Using and Aimed Throwing in a Community of Free Living Chimpanzees." *Nature* 201: 1264–1266.

GOODMAN, M. E. (1964) *Race Awareness in Young Children.* New York: Crowell-Collier Books.

———. (1967) *The Individual and Culture.* Homewood, Ill.: Dorsey.

GOSLIN, D. A. (ed.). (1969) *Handbook of Socialization Theory and Research.* Chicago: Rand McNally.

GRAVEN, J. (1967) *Non-Human Thought.* New York: Stein and Day.

GRANQVIST, H. (1947) *Birth and Childhood Among the Arabs.* Helsinki: Soderstrom.

———. (1950) *Child Problems Among the Arabs.* Helsinki: Soderstrom.

GREENBERG, J. (1963) *Universals of Language.* Cambridge, Mass.: MIT Press.

GREENBAUM, L. (1970) "Evaluation of a Stratified Versus an Unstratified Universe of Cultures in Comparative Research." *Behavior Science Notes* 5: 251–289.

GREENSTEIN, F. I. (1968) "Socialization: Political Socialization." In W. A. Wallis (ed.), *International Encyclopedia of the Social Sciences.* New York: Macmillan and the Free Press, vol. 19, pp. 551–555.

GRINNELL, G. B. (1923) *The Cheyenne Indians: Their History and Ways of Life.* New Haven: Yale University Press.

GUTHRIE, G. M. and P. J. JACOBS. (1966) *Child Rearing and Personality Development in the Philippines.* University Park: Pennsylvania State University Press.

HALL, E. T. (1964) "Adumbration as a Feature of Intercultural Communication." *American Anthropologist* 66: 154–163.

HALLOWELL, A. I. (1950) "Personality Structure and the Evolution of Man." *American Anthropologist* 52: 159–173.

———. (1953) "Culture, Personality and Society." In A. L. Kroeber (ed.), *Anthropology Today.* Chicago: University of Chicago Press, pp. 597–620.
———. (1954a) "The Self and Its Behavioural Environment." *Explorations II* (April), 106–165.
———. (1954b) "Psychology and Anthropology." In J. Gillin (ed.), *For a Science of Social Man.* New York: Macmillan, pp. 160–226.
———. (1956) "The Structural and Functional Dimensions of a Human Existence." *Quarterly Review of Biology* 31: 88–101.
———. (1959) "Behavioral Evolution and the Emergence of the Self." In B. J. Meggars (ed.), *Evolution and Anthropology: A Centennial Appraisal.* Washington, D.C.: Anthropological Society of Washington.
———. (1960) "Self, Society and Culture in Phylogenetic Perspective." In Sol Tax (ed.), *Evolution After Darwin:* vol. 2, *The Evolution of Man.* Chicago: University of Chicago Press, pp. 309–371.
———. (1961) "The Protocultural Foundations of Human Adaptation." In S. L. Washburn (ed.), *Social Life of Early Man.* Chicago: Aldine, pp. 236–255.
———. (1963) "Personality, Culture and Society in Behavioral Evolution." In S. Koch (ed.), *Psychology: A Study of a Science.* New York: McGraw-Hill, vol. 6, pp. 429–509.
———. (1965) "Hominid Evolution, Cultural Adaptation and Mental Dysfunctioning." In A. V. S. de Reuck and R. Porter (eds.), *Transcultural Psychiatry.* Boston: Little, Brown, pp. 26–54.
———. (1967) *Culture and Experience.* New York: Schocken Books (originally published 1955, Philadelphia: University of Pennsylvania Press).
HARING, D. G. (1947) "Science and Social Phenomena." *American Scientist* 35: 351–363.
———. (1956) *Personal Character and Cultural Milieu,* 1st rev. ed. Syracuse, N.Y.: Syracuse University Press.
HEINICKE, C. and B. B. WHITING. (1953) *Bibliography on Personality and Social Development of the Child and Selected Ethnographic Sources on Child Training.* New York: Social Science Research Council, Pamphlet Number 10.
HENRY, J. (1960) "A Cross-Cultural Outline of Education." *Current Anthropology* 1: 267–305.
HILGER, SISTER M. I. (1960) *Field Guide to the Ethnological Study of Child Life.* New Haven: Human Relations Area Files.
HOCKETT, C. (1958) *A Course in Modern Linguistics.* New York: Macmillan.
——— and R. ASCHER (1964) "The Human Revolution." *Current Anthropology* 5: 135–168.
HOGBIN, H. I. (1930) "Spirits and the Healing of the Sick in Ontong-Java." *Oceania* 1: 146–166.
———. (1931) "Education at Ontong-Java." *American Anthropologist* 33: 601–614.
———. (1943) "A New Guinea Infancy: From Conception to Weaning in Wogeo." *Oceania* 13: 235–309.
———. (1946) "A New Guinea Infancy: From Weaning to the Eighth Year in Wogeo." *Oceania* 16: 275–296.

HOMANS, G. C. (1950) *The Human Group.* New York: Harcourt Brace Jovanovich.

———. (1961) *Social Behavior: Its Elementary Forms.* New York: Harcourt Brace Jovanovich.

HONIG, W. K. (ed.). (1966) *Operant Behavior: Areas of Research and Application.* New York: Appleton-Century-Crofts.

HONIGMANN, J. (1954) *Culture and Personality.* New York: Harper & Row.

———. (1967) *Personality in Culture.* New York: Harper & Row.

HOPPE, R., G. MILTON, and E. SIMMEL (eds.) (1970) *Early Experiences and the Processes of Socialization.* New York: Academic Press.

HOWELL, F. C. (1969) "Remains of Hominids from Pliocene/Pleistocene Formations in the Lower Omo Basin, Ethiopia." *Nature* 223: 1234–1239.

HSU, F. L. K. (1961) *Psychological Anthropology.* Homewood, Ill.: Dorsey.

HUNT, R. (ed.) (1967) *Personalities and Cultures.* Garden City, N.Y.: Natural History Press.

HYDÉN, H. and E. EGYHÁZI. (1964) "Changes in RNA Content and Base Composition in Cortical Neurons of Rats in a Learning Experiment Involving Transfer of Handedness." *Proceedings of the National Academy of Sciences, U.S.A.,* 52: 1030–1035.

HYMES, D. H. (ed.). (1964) *Language in Culture and Society.* New York: Harper & Row.

ITANI, J. (1959) "Paternal Care in the Wild Japanese Monkey." *Primates* 4: 11–66.

JAKOBSON, R. and M. HALLE. (1966) *Fundamentals of Language.* The Hague: Mouton.

JAY, P. C. (1968) *Primates: Studies in Adaptation and Variability.* New York: Holt, Rinehart and Winston.

JOSEPH, A., R. SPICER, and J. CHESKY. (1949) *The Desert People.* Chicago: University of Chicago Press.

KAPLAN, B. (ed.). (1961) *Studying Personality Cross-Culturally.* Evanston, Ill.: Row, Peterson.

KAY, B. (1957) "The Reliability of HRAF Coding Procedures." *American Anthropologist* 59: 524–527.

KIDD, D. (1906) *Savage Childhood: A Study of Kafir Children.* London: Macmillan and A. and C. Black.

KLOPP, O. E. (1964) "Mexican Social Types." *American Journal of Sociology* 69: 404–414.

KLUCKHOHN, C., H. A. MURRAY, and D. SCHNEIDER (eds.). (1956) *Personality in Nature, Society and Culture,* 2nd rev. ed. New York: Knopf.

KROEBER, A. L. (1948) *Anthropology.* New York: Harcourt, Brace.

——— and C. KLUCKHOHN. (1952) *Culture: A Critical Review of Concepts and Definitions.* Papers of the Peabody Museum of American Archaeology and Ethnology, Harvard University, XLVII.

KUHN, T. B. (1966) *The Structure of Scientific Revolutions.* Chicago: University of Chicago Press.

KUMMER, H. (1968) "Two Variations in the Social Organization of Baboons." In P. C. Jay (ed.), *Primates: Studies in Adaptation and Variability.* New York: Holt, Rinehart and Winston, pp. 293–312.

LA BARRE, W. (1954) *The Human Animal.* Chicago: University of Chicago Press.

LANCASTER, J. B. (1967) *Primate Communication Systems and the Emergence of Human Language.* Unpublished Ph.D. dissertation.

LANDY, D. (1959) *Tropical Childhood.* Chapel Hill: University of North Carolina Press.

LANTIS, M. (1960) *Eskimo Childhood and Interpersonal Relationships.* Seattle: University of Washington Press.

LEAKEY, R. E. F., A. K. BEHRENSMEYER, F. J. FITCH, J. A. MILLER, and M. D. LEAKEY. (1970) "New Hominid Remains and Early Artifacts from Northern Kenya." *Nature* 226: 223–230.

LEGROS CLARK, W. E. (1962) *The Antecedents of Man.* Edinburgh: University of Edinburgh Press.

LEIGHTON, D. and C. KLUCKHOHN. (1947) *Children of the People.* Cambridge, Mass.: Harvard University Press.

LEIS, P. E. (1964) "Ijaw Enculturation: A Reexamination of the Early Learning Hypotheses." *Southwestern Journal of Anthropology* 20: 32–42.

LENNEBERG, E. (1964) *New Directions in the Study of Language.* Cambridge, Mass.: M.I.T. Press.

———. (1967) *Biological Foundations of Language.* New York: Wiley.

LEVINE, R. A. (1969) "Culture, Personality and Socialization: An Evolutionary View." In D. A. Goslin (ed.), *Handbook of Socialization Theory and Research.* Chicago: Rand McNally, pp. 503–541.

LEVY, M. J. (1952) *The Structure of Society.* Princeton, N.J.: Princeton University Press.

LEWIS, O. (1951) *Life in a Mexican Village: Tepotzlan Restudied.* Urbana: University of Illinois Press.

LINDESMITH, A. R. and A. L. STRAUSS. (1950) "A Critique of Culture—Personality Writings." *American Sociological Review* 15: 587–600.

MACGREGOR, G. (1946) *Warriors Without Weapons.* Chicago: University of Chicago Press.

MCNEIL, E. (1969) *Human Socialization.* Belmont, Calif.: Brooks/Cole.

MCNEILL, D. (1970) *The Acquisition of Language: The Study of Developmental Psycholinguistics.* New York: Harper & Row.

MARTIN, W. (1960) "Rediscovering the Mind of the Child: A Significant Trend in Research in Child Development." *Merrill-Palmer Quarterly* 6: 67–76.

MASLOW, A. H. (1954) *Motivation and Personality.* New York: Harper.

MATHEWS, R. (1907) "Notes on the Aranda Tribe." *Journal and Proceedings of the Royal Society of New South Wales,* Number 41.

———. (1908) "Marriage and Descent in the Aranda Tribe, Central Australia." *American Anthropologist* 10: 88–102.

MAYER, P. (ed.). (1970) *Socialization: The Approach from Social Anthropology.* London: Tavistock.

MAYR, E. (1958) "Behavior and Systematics." In A. Roe and G. C. Simpson (eds.), *Behavior and Evolution.* New Haven, Conn.: Yale University Press, pp. 341–362.

———. (1963) *Animal Species and Evolution.* Cambridge, Mass.: Harvard University Press.

MEAD, M. (1928) *Coming of Age in Samoa.* New York: Morrow (1961, New York: Morrow Apollo Edition).
———. (1930) *Growing Up in New Guinea.* New York: Morrow.
———. (1935) *Sex and Temperament in Three Primitive Societies.* New York: Morrow.
———. (1937) *Cooperation and Competition Among Primitive Peoples.* New York: McGraw-Hill.
———. (1947) "On the Implications for Anthropology of the Gesell-Ilg Approach to Maturation." *American Anthropologist* 49: 69–77.
———. (1949a) *Male and Female: A Study of the Sexes in a Changing World.* New York: Morrow.
———. (1949b) "Psychologic Weaning: Childhood and Adolescence." In P. Hoch and J. Zubin (eds.), *Psychosexual Development in Health and Disease.* New York: Grune and Stratton, pp. 124–135.
———. (1956a) "The Concept of Culture and the Psychosomatic Approach." In D. G. Haring (ed.), *Personal Character and the Cultural Milieu.* Syracuse, N.Y.: Syracuse University Press, pp. 594–662.
———. (1956b) *New Lives for Old: Cultural Transformation-Manus.* 1928–1953, New York: Morrow.
———. (1957) "Towards More Vivid Utopias." *Science* 126: 957–961.
———. (1959) "Closing the Gap Between Scientists and Others." *Daedalus* (Winter, 1959), 139–146.
———. (1961) "Cultural Determinants of Sexual Behavior." In W. C. Young (ed.), *Sex and Cultural Secretions.* 3rd ed. Baltimore: Williams and Wilkins, vol. 2, pp. 1433–1479.
———. (1963) "Socialization and Enculturation." *Current Anthropology* 4: 184–188.
———. (1965) "The Future as the Basis for Establishing a Shared Culture." *Daedalus* (Winter 1965), 135–155.
———. (1970) *Culture and Commitment: A Study of the Generation Gap.* New York: Natural History Press/Doubleday.
———. (1971) "Options Implicit in Developmental Styles." In E. Tobach (ed.), *The Biopsychology of Development.* New York: Academic Press, pp. 533–541.
——— and F. C. MACGREGOR. (1951) *Growth and Culture: A Photographic Study of Balinese Childhood.* New York: Putnam's.
——— and M. WOLFENSTEIN (eds.). (1955) *Childhood in Contemporary Cultures.* Chicago: University of Chicago Press.
——— and N. NEWTON. (1967) "Cultural Patterning of Perinatal Behavior." In A. F. Guttmacher and S. Richardson (eds.), *Childbearing: Its Social and Psychological Aspects.* Baltimore: Williams and Wilkins, pp. 142–245.
MENAKER, E. and W. MENAKER. (1965) *Ego in Evolution.* New York: Grove.
MERTON, R. K. (1957) *Social Theory and Social Structure,* rev. ed. New York: Free Press.
MILLER, N. (1969) "Learning of Visceral and Glandular Responses." *Science* 163: 434–444.
——— and L. DICARA. (1968) "Instrumental Learning of Systolic Blood Pres-

sure Responses by Curarized Rats: Dissociation of Cardiac and Vascular Changes." *Psychosomatic Medicine* 30: 489–494.

MILNER, R. and J. PROST. (1967) "The Significance of Primate Behavior for Anthropology." In N. Korn and F. Thompson (eds.), *Human Evolution: Readings in Physical Anthropology.* New York: Holt, Rinehart and Winston, pp. 125–136.

MIYADI, D. (1964) "Social Life of Japanese Monkeys." *Science* 143: 783–786.

MONTAGU, M. F. A. (ed.). (1962) *Culture and the Evolution of Man.* New York: Oxford University Press.

MURDOCK, G. P. (1949) *Social Structure.* New York: Macmillan.

———. (1958) *Outline of World Cultures,* 2nd rev. ed. New Haven, Conn.: Human Relations Area Files.

———. (1967) *Ethnographic Atlas.* Pittsburgh: University of Pittsburgh Press.

——— *et al.* (1961) *Outline of Cultural Materials.* New Haven, Conn.: Human Relations Area Files.

NAPIER, J. (1970) "The Roots of Mankind." *The Listener* (March 12, 1970), pp. 342–344.

NAROLL, R. (1961) "Two Solutions to Galton's Problem." *Philosophy of Science* 23: 15–39.

———. (1964) "On Ethnic Unit Classification." *Current Anthropology* 5: 283–291.

———. (1968) "Some Thoughts on Comparative Method in Anthropology." In H. Blalock and A. Blalock (eds.), *Sociological Method.* New York: McGraw-Hill, pp. 236–277.

———. (1970) "What Have We Learned from Cross-Cultural Surveys?" *American Anthropologist* 72: 1227–1288.

——— and R. D'ANDRADE. (1963) "Two Further Solutions to Galton's Problem." *American Anthropologist* 65: 1053–1067.

———, W. ALNOT, J. CAPLAN, J. HANSEN, J. MAXANT, and N. SCHMIDT. (1970) "A Standard Ethnographic Sample: Preliminary Edition." *Current Anthropology* 11: 235–248.

NEW YORK TIMES. (1970) "New Clue Found on Man and Ape." *The New York Times* (November 29, 1970), p. 65.

OAKLEY, K. P. (1968) "The Earliest Tool-Makers." In G. Karth (ed.), *Evolution and Hominisation,* 2nd ed. Stuttgart: Gustav Fisher Verlag, pp. 257–272.

———. (1969) "Man the Skilled Tool Maker." *Antiquity* 43: 222–224.

———. (1970) "Pliocene Men." *Antiquity* 44: 307–308.

O'LEARY, T. J. (1969) "A Preliminary Bibliography of Cross-Cultural Studies." *Behavior Science Notes* 4: 95–115.

OPLER, M. E. (1941) *An Apache Life Way.* Chicago: University of Chicago Press.

ORLANSKY, H. (1949) "Infant Care and Personality." *Psychological Bulletin* 46: 1–48.

PARSONS, T. (1949) *The Structure of Social Action,* 2nd ed. New York: Free Press.

———. (1951) *The Social System.* New York: Free Press.
——— and R. F. BALES. (1955) *Family, Socialization and Interaction Process.* New York: Free Press.
——— and E. A. SHILS. (1951) *Toward a General Theory of Action.* Cambridge, Mass.: Harvard University Press.
PIAGET, J. (1929) *The Child's Conception of the World.* New York: Humanities.
———. (1947) *Judgment and Reasoning in the Child.* New York: Humanities.
———. (1954) *Construction of Reality in the Child.* New York: Basic Books.
———. (1955) *Language and Thought of the Child.* Cleveland: World.
——— and B. INHELDER. (1948) *The Child's Conception of Space.* New York: Humanities.
PINK, O. (1936) "The Landowners of the Northern Division of the Aranda Tribe. Central Australia." *Oceania* 6.
RABIN, A. I. (1965) *Growing Up in the Kibbutz.* New York: Springer.
RADCLIFFE-BROWN, A. R. (1930–1931) "The Social Organization of Australian Tribes." *Oceania* 1.
RAUM, O. F. (1940) *Chagga Childhood.* London: Oxford University Press.
READ, M. (1960) *Children of Their Fathers: Growing Up Among the Ngoni of Nyasaland.* New Haven, Conn.: Yale University Press.
———. (1968) *Growing Up Among the Ngoni of Malawi.* New York: Holt, Rinehart and Winston.
RICHARDS, A. I. (1956) *Chisungu.* New York: Grove.
RICHIE, G. E. (1963) *The Making of a Maori.* Wellington, New Zealand: A. H. and A. W. Reed.
ROE, A. and G. G. SIMPSON (eds.). (1958) *Behavior and Evolution.* New Haven, Conn.: Yale University Press.
SAHLINS, M. D. (1959) "The Social Life of Monkeys, Apes and Primitive Man." *Human Biology* 31: 54–73.
———. (1960) "The Origin of Society." *Scientific American* 203: 76–87.
SCHRIER, A. M., H. F. HARLOW, and F. STOLLNITZ. (1965) *Behavior of Nonhuman Primates.* New York: Academic Press.
SEARS, R. R., E. E. MACCOBY, and H. LEVIN. (1957) *Patterns of Child Rearing.* Evanston, Ill.: Row, Peterson.
SEBEOK, T. A. (ed.). (1968) *Animal Communication; Techniques of Study and Results of Research.* Bloomington, Ind.: University of Indiana Press.
——— and A. RAMSEY (eds.). (1969) *Approaches to Animal Communication.* The Hague: Mouton.
SELIGMAN, B. Z. (ed.). (1951) *Notes and Queries on Anthropology: Sixth Edition, Revised and Rewritten by a Committee of the Royal Anthropological Institution of Great Britain and Ireland.* London: Routledge and Kegan Paul.
SEWELL, W. H. (1952) "Infant Training and the Personality of the Child." *American Journal of Sociology* 58: 150–159.
———. (1963) "Some Recent Developments in Socialization Theory and Research." *Annals of the American Academy of Political and Social Science* 349: 163–181.

SHERIF, M. (1935) "A Study of Some Social Factors in Perception." *Archives of Psychology* 27.

SHIMAHARA, N. (1970) "Enculturation—A Reconsideration." *Current Anthropology* 11: 143–154.

SKINNER, B. F. (1938) *The Behavior of Organisms.* New York: Appleton-Century-Crofts.

SOUTHWICK, C. H. (ed.). (1963) *Primate Social Behavior.* New York: Van Nostrand.

SPENCER, B. (1928) *Wanderings in Wild Australia.* London: Macmillan.

———. (1938) *The Native Tribes of Central Australia.* London: Macmillan (2 vols.).

——— and F. GILLEN. (1927) *The Arunta: A Stone Age People.* London: Macmillan.

SPIRO, M. E. (1951) "Culture and Personality: The Natural History of a False Dichotomy." *Psychiatry* 14: 19–46.

———. (1954) "Human Nature in Its Psychological Dimensions." *American Anthropologist* 56: 19–30.

———. (1958) *Children of the Kibbutz.* Cambridge, Mass.: Harvard University Press.

——— (ed.). (1965) *Context and Meaning in Cultural Anthropology.* New York: Harper.

SPUHLER, J. N. (1959) "Somatic Paths to Culture." *Human Biology* 31: 1–13.

STOCK, R. (1969) "The Mouse State of the New Biology." *New York Times Magazine* (December 21, 1969), pp. 8 ff.

STREHLOW, T. G. H. (1947) *Aranda Traditions.* Melbourne: Melbourne University Press.

TEXTOR, R. (1967) *A Cross-Cultural Summary.* New Haven, Conn.: HRAF Press.

TIGER, L. (1969) *Men in Groups.* New York: Random House.

WADDINGTON, C. H. (1960) *The Ethical Animal.* London: G. Allen and Unwin.

WALLACE, A. F. C. (1961) *Culture and Personality.* New York: Random House (1970, rev. ed.).

WASHBURN, S. L. (1950) "The Analysis of Primate Evolution with Particular Reference to the Origin of Man." *Cold Spring Harbor Symposia on Quantitative Biology* 15: 67–77.

———. (1958) "Speculation on the Interrelations of the History of Tools and Biological Evolution." *Human Biology* 31: 21–31.

———. (1963) "Behavior and Human Evolution." In S. L. Washburn (ed.), *Classification and Human Evolution.* Chicago: Aldine, pp. 190–203.

——— and I. DEVORE. (1961) "The Social Life of Baboons." *Scientific American* 204: 62–71.

———, P. C. JAY, and J. B. LANCASTER. (1965) "Field Studies of Old World Monkeys and Apes." *Science* 150: 1541–1547.

WHITING, J. W. M. (1941) *Becoming a Kwoma.* New Haven, Conn.: Yale University Press.

———. (1968) "Socialization: Anthropological Aspects." In W. A. Wallis (ed.), *International Encyclopedia of the Social Sciences.* New York:

Macmillan and the Free Press, vol. 14, pp. 545–551.
——— and I. L. CHILD. (1953) *Child Training and Personality.* New Haven, Conn.: Yale University Press.
WIENER, N. (1950) "Some Maxims for Biologists and Psychologists." *Dialectica* 4: 22–27.
WILLIAMS, T. R. (1958) "The Structure of the Socialization Process in Papago Indian Society." *Social Forces* 36: 251–256. (Reprinted in J. Middleton [ed.], *From Child to Adult: Studies in Anthropology of Education.* New York: Natural History Press, 1970, pp. 163–172.)
———. (1959) "The Evolution of a Human Nature." *Philosophy of Science,* 26: 1–13.
———. (1966) "Cultural Structuring of Tactile Experience in a Borneo Society." *American Anthropologist* 68: 27–34.
———. (1969) *A Borneo Childhood: Enculturation in Dusun Society.* New York: Holt, Rinehart and Winston.
———. (1972) *Introduction to Socialization: Human Culture Transmitted.* St. Louis: Mosby.
WILSON, M. (1951) *Good Company.* London: Oxford University Press.